FutureLife

The Biotechnology Revolution

Dr. Alvin Silverstein and Virginia B. Silverstein

Illustrated with photographs
Line drawings by Marjorie Thier

Prentice-Hall, Inc.
Englewood Cliffs, New Jersey

For Barbara Francis and Carol Barkin

Printed in the United States of America   ·J

Prentice-Hall International, Inc., London
Prentice-Hall of Australia, Pty. Ltd., Sydney
Prentice-Hall of Canada, Ltd., Toronto
Prentice-Hall of India Private Ltd., New Delhi
Prentice-Hall of Japan, Inc., Tokyo
Prentice-Hall of Southeast Asia Pte. Ltd., Singapore
Whitehall Books Limited, Wellington, New Zealand

10 9 8 7 6 5 4 3 2 1

Library of Congress Cataloging in Publication Data

Silverstein, Alvin
    Futurelife, the biotechnology revolution.

    Bibliography: p.
    Includes index.
    Summary: Discusses new technologies which have the
potential for wiping out disease, breeding new species of
plants and animals, manufacturing exotic chemicals, and
improving the quality of life.
    1. Bioengineering—Juvenile literature.   2. Genetic
engineering—Juvenile literature.   [1. Bioengineering.
2. Genetic engineering.   3. Technology]   I. Silverstein,
Virginia B.   II. Title.
TA164.S54   600   81-21104
ISBN 0-13-345884-9   AACR2

ACKNOWLEDGMENTS

The authors are grateful to all those who so generously supplied photographs and reference materials for the book. We wish that space had permitted all the excellent photos to be used. Special thanks to Alastair Fraser of the Hunterdon Medical Center, Flemington, New Jersey, for providing some fascinating glimpses of biotechnology in action.

CONTENTS

THE BIOTECHNOLOGY REVOLUTION

The revolution has begun. The war is being fought all over the world. The battlefields are in the laboratories and hospitals. The fighters are scientists and doctors. Their goal is to save, improve, and to extend life in ways considered impossible until now. Their enemy is the unknown.

This is a biomedical revolution. Technology is providing the weapons for the fight. New tools are permitting us to see into the body in ways that were never possible before. With these devices, doctors are getting an early start at treating people who are ill. And the earlier problems are found, the easier they are to solve.

Technology is bringing new ways of treating diseases. Lasers to burn out cancers, artificial kidneys and hearts to take over the work of organs that have failed—these are part of the biotechnology revolution that is changing our lives.

Another part of the fight is being waged in test tubes and culture dishes. Researchers are probing down to the microscopic cells of the body to find out what makes them go wrong. Teams of scientists are breeding armies of microbes to defend and serve humankind. Genetic engineers are changing heredity. They are learning to shape the living world to their desires. Some day we may even have the skills to remake ourselves.

With the new tools and skills of biotechnology come new problems, new responsibilities. Decisions need to be made about how we should use the new technologies. Not only scientists but also ordinary citizens need to know and understand the biotechnology revolution that is sweeping through our world, so that the decisions we make will be wise ones.

New Tools
For
Medicine

A hundred years ago when people were ill, doctors did not have many ways to find out what was going wrong inside them. There were stethoscopes for listening to the sounds of the heart and lungs. Doctors could look at their patients' skin, feel for any suspicious bumps or lumps, and examine samples of their urine, saliva, and other body products. Tools had been invented for peering into a patient's eyes or down a throat. But doctors couldn't see what was happening inside their patients' bodies without actually cutting them open—and that was very dangerous.

In 1895 a German physicist, Wilhelm Roentgen, took a picture of his wife that started a revolution in medicine. It was an unusual picture. All it showed were the bones of Mrs. Roentgen's hand, with her gold wedding

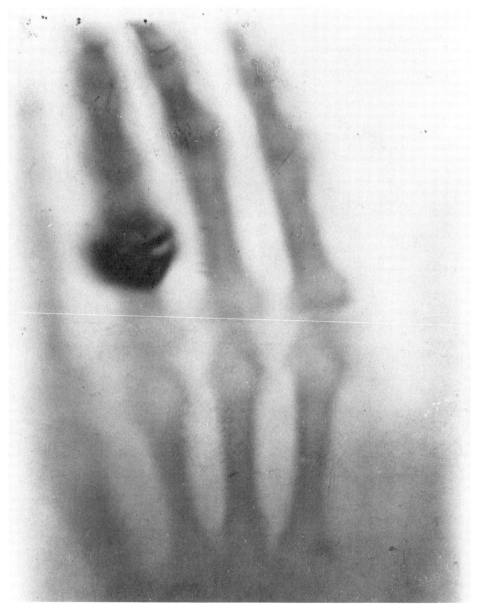

The first X-ray ever made showed the bones of Mrs. Roentgen's hand.

4

ring eerily suspended on one fingerbone. Instead of using ordinary light to take the picture, Roentgen used X-rays. Unlike ordinary light rays, X-rays can pass right through solid objects. But they pass through soft tissues of the body much more easily than through hard bone. So the bones of the skeleton show up as shadows. (The X-ray prints that doctors use today are negatives, so that bones show up white against a black background.)

Newspapers all over the world published the picture of Mrs. Roentgen's hand, and X-ray pictures quickly became the latest fad. Fashionable ladies had X-ray pictures taken of their hands while they were wearing their fanciest rings. Engaged couples sentimentally treasured X-ray images of their hands clasped together. Meanwhile, doctors discovered that X-ray pictures permitted them to find gallstones and to locate foreign objects lodged in their patients' flesh. An X-ray of a broken bone showed exactly where the fracture was; after a cast was put on, another X-ray could be used to check that the bone had been set properly.

As the years went by, medical researchers found that too much exposure to X-rays could be dangerous. But X-ray machines were improved again and again, permitting doctors to take much sharper pictures in much shorter times. (In 1896 a person had to stay still under the X-ray beam for at least twenty minutes to take a picture. Now an X-ray picture can be taken in a fraction of a second.) Doctors even have ways to make soft tissues and organs of the body show up on X-ray photos. The patient may take pills containing a special dye or drink a chalky-tasting solution of a barium salt. These substances outline the body organs, so that they show up just as clearly as bones.

The use of X-rays has become so commonplace in medical practice that we take it for granted. Recently, researchers have developed a number of new tools for looking at and into the body, which are bringing just as revolutionary and exciting changes in medicine as the first X-ray pictures.

One of these new tools is the CAT scanner. It doesn't have anything to do with cats; its name is an abbreviation for computerized axial tomography. That is a fancy way of saying that a CAT scanner uses a computer to build up a vivid image of a "slice" of a person's head or body. The computer puts together a large number of X-ray images taken at different angles by an

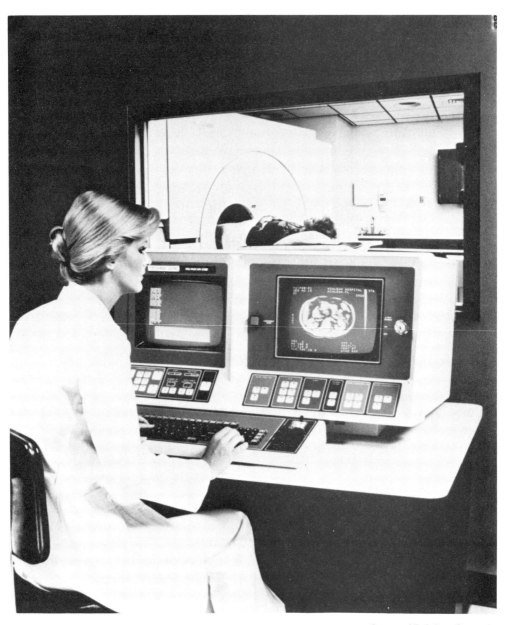

A CAT body scanner in operation.

X-ray beam traveling in a circle around the long axis of the person's body. The machine that takes the pictures looks like a big doughnut. The patient, lying on a table, passes into the hole in the "doughnut."

The first CAT scanners, introduced in 1972, were only big enough to take pictures of a patient's head, and they were very expensive: a single machine cost about $350,000. But they were very useful. If a person had a head injury, for example, doctors could use a CAT scan to find out if there was bleeding inside the brain and exactly where the problem was. People who were troubled by headaches could learn whether they had a brain tumor or cyst. Surgeons, knowing before they operated exactly where the tumor was and how large it was, could plan their operations so they would not hurt the healthy brain tissue.

A few years later, body scanners were developed. They were even more expensive than the brain scanners, but doctors and hospitals all over the world eagerly ordered and used them. Before CAT scanners, doctors often had to cut open a patient's body to find out what was going on inside. But now they could look at a "slice" of the patient's organs pictured on a TV screen. Tumors were outlined clearly. The CAT scans could even show whether a tumor was cancerous or not. A doctor could drain a cyst in a kidney or pancreas without surgery by watching carefully on a CAT scan while inserting a hollow needle. CAT scans were also used after surgery to detect infections before they became serious. Doctors used this new tool in follow-up studies, to see how well their treatments were working.

The use of CAT scanners grew rapidly. After awhile, government agencies and consumer groups began to ask worried questions. Were all these CAT scanners necessary? Maybe they were just a big waste of money. Maybe they were even dangerous. Orders for new CAT scanners were slowed down, and careful studies were made. The studies showed that a small hospital probably would not get enough use from a CAT scanner to make it worthwhile; but in large hospitals, the new tool saves money. And wherever it is used, it saves lives. The X-ray exposure from a CAT scan is not much greater than from ordinary X-ray pictures, and this test can often be used instead of other, more dangerous tests. Sometimes a CAT scan provides information that the doctor could not get otherwise, except by operating. In

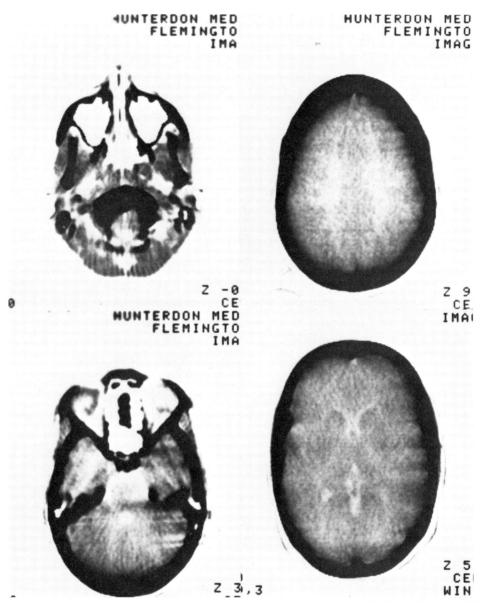

*These CAT scans, taken at different levels (from the bottom upward),
show images of "slices" of a normal brain.*

about a quarter of the cases, the results of a CAT scan show that an operation is not needed. Surgical operations are very expensive, and even with the most skillful surgeons, some of the patients die. The CAT scanner has turned out to be so valuable that its inventors, Allan Cormack and Godfrey Hounsfield, received a Nobel Prize in 1979.

In 1980 researchers at the Mayo Clinic reported on an even more exciting X-ray scanner, the DSR (dynamic spatial reconstructor). The DSR uses a computer and X-rays to build up a vivid three-dimensional image of the body in action. Using a DSR, for example, researchers can watch a living heart beating. The electronic image on a TV screen can be rotated, turned over, or even opened up to see what is happening inside. The computer can electronically ''dissolve away'' the muscle tissue to show the arteries that carry blood to the heart. Doctors can watch blood flowing through the heart or air moving in and out of the lungs. And all this can be done without harming the body at all. The new scanner will be a great help in treating patients with heart and lung problems. Like a CAT scanner, it can show up damage in a heart wall or a plug in an artery. But it gives a picture of the whole organ, not just a slice of it, and it can also show up problems in the way the organ works.

Another kind of scanner that is bringing doctors and medical researchers a wealth of new information is the PET scanner. PET is an abbreviation for positron emission tomography. Before taking a PET scan, the patient is injected with a sugar solution ''tagged'' with a special radioactive substance. The tagged sugar is carried in the bloodstream to various parts of the body. Along the way, bits of the radioactive substance are constantly breaking down and producing (''emitting'') minute particles called positrons. Positrons are very unstable in our world. As soon as one meets an electron, they combine explosively and shoot out a kind of radiation called gamma rays. Since all the atoms in our world contain electrons, a positron gets destroyed practically as soon as it is made. It sounds a little scary for all that to be going on inside a person's body, but the amounts of radioactivity involved are too tiny to hurt anybody. Special gamma ray detectors can pick up the traces of positrons, and these traces are used to build up the picture in a PET scanner.

The DSR scanner produces three-dimensional images that can be viewed from any angle, as in these rotated views of the monkey's lungs, and even "dissected" electronically.

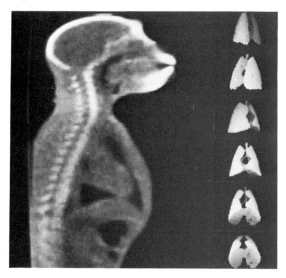

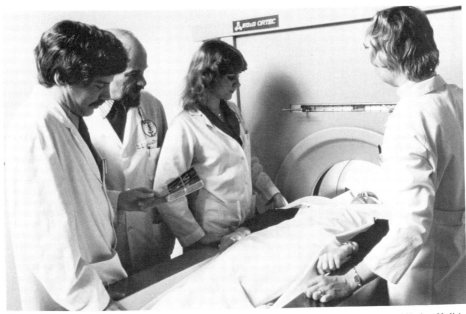

National Institutes of Health Clinical Center, Department of Nuclear Medicine

The PET scanner uses radioactively labeled chemicals to map the active areas of tissues and organs.

A CAT scanner gives a picture of structures inside the body. A PET scanner does too, but its images show which structures are most actively using sugar. Since sugar is the main energy food, especially in the brain, a PET scan shows which cells are working actively. Researchers can watch the "hot spots" that show up in various parts of the brain when a person sees a light or listens to a voice or moves an arm or leg. PET scans can show doctors how best to treat people who have had strokes due to blocked arteries in the brain, and they can reveal where epileptic seizures start. PET scans can also help to diagnose cancer, since many tumors take up sugar faster than normal tissues.

Studies of people with mental illness have yielded striking results. People with schizophrenia show special patterns of brain activity in a PET scan. Manic-depressives show another pattern. These patterns are quite different from the brain patterns of normal people. PET scans are making it much

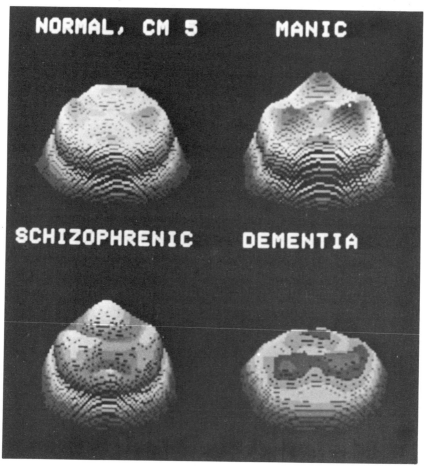

PET scans are providing new insight into the way the body and brain work. These PET scans show contrasting patterns of brain activity in a normal person and in patients suffering from various kinds of mental illness.

easier and quicker to diagnose certain mental disorders and may help doctors in treating them. Unfortunately, PET scanners are even more expensive than CAT scanners. In addition to the PET equipment, a machine called a cyclotron is needed to make the special radioactive compounds. The cost of PET scanning equipment can come to millions of dollars.

Scanners that work by sending X-rays or other radiation through the body carry a risk of harming the patient. It is a small risk, but doctors would prefer to have no risk at all—or at least the smallest risk possible. Now medical researchers are excited about a new kind of scanner that is being developed, the NMR scanner. NMR stands for nuclear magnetic resonance. It is based on the fact that in a field of radio waves, the nuclei of certain atoms act like tiny magnets and line up all together. If another radio field is turned on at right angles to the first, the nuclei start to spin like miniature tops. They wobble as they spin, and after awhile they fall back to their original position. The spinning nuclei send out signals that can be picked up, and the signals and the times it takes for the nuclei to fall back into place are information that a computer can use to make a picture.

Hydrogen nuclei send out the strongest NMR signals, and hydrogen atoms are very plentiful in the body. They are especially plentiful in water molecules, H_2O. So the areas that show up strongest in NMR scans are areas with a lot of water. These scans are good for observing blood flows and, especially, for picking out internal bleeding. NMR scans can also show tumors, because tumor cells apparently do not hold their water molecules as tightly as normal cells do. In one laboratory where researchers were testing a new NMR scanner, they were taking scans of people with brain tumors and blood vessel disorders and comparing them with CAT scans. A visitor to the lab asked to have an NMR scan done to see what it was like. Just as the researchers had assured him, the man found that having an NMR scan didn't hurt. But he was surprised to hear the doctors exclaiming as they looked at the picture of his brain. The scan showed something that should not be there: a growth just behind the man's eye. Suddenly the visitor became a patient. He was sent for a CAT scan and other tests. They showed that the NMR scan was correct. The man had a tumor. Before his visit to the lab, he had been suffering from severe headaches, but no one had realized what was wrong. Surgeons removed the tumor. After the operation, the headaches were gone.

The NMR scanner does not subject a patient to any radiations, but it does use radio waves. Researchers worried at first that the radio waves might cook the tissues, acting like a microwave oven. But tests of radio waves

NMR body scanner: the NMR scans are somewhat similar to the images produced by a CAT scanner.

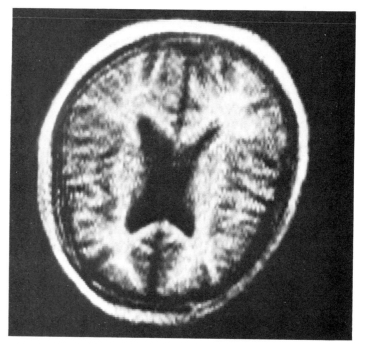

NMR scanners use signals from tiny spinning nuclei of atoms in the body to build up images of "slices."

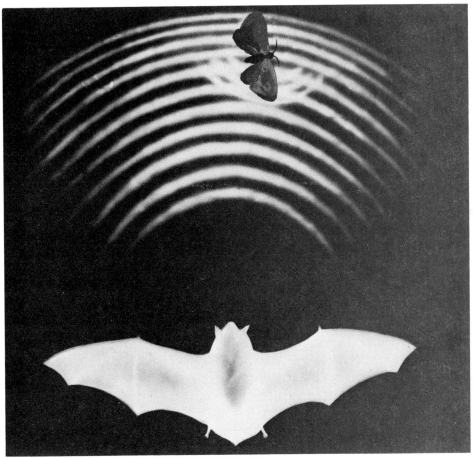

The bat locates its prey in the dark by bouncing ultrasound waves off objects.

at the frequencies used in NMR scanners have not shown any damage to the body or its tissues.

Another kind of scanning that seems particularly safe for the patient is ultrasonography. Pictures are formed by bouncing ultrasound—very high-pitched sound waves—off body structures. (These sound waves are too high-pitched for human ears to hear.) Bats use a kind of natural ultrasonography to find their prey and fly around in the dark. They send out high-pitched

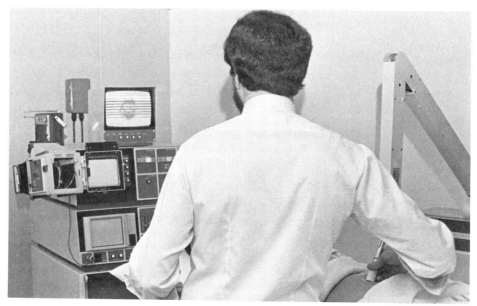

The probe the technician is placing against the patient's abdomen sends out ultrasound waves that bounce off internal structures to form a picture.

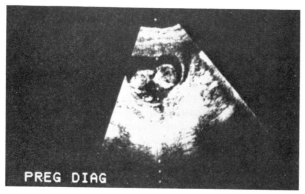

An ultrasonic scan of a 12-week fetus. The fetus' head (with eyes), body, and limbs can be seen inside its mother's womb. The large mass above it is the placenta, which provides nourishment.

16

cries of ultrasound waves that bounce off objects. A large part of the bat's brain is like a specially adapted computer, which analyzes the way the sound waves bounce and builds their patterns into a picture of tree branches, fluttering moths, and other things in the bat's path. Dolphins and whales use a similar ultrasonic location system to find their way through the underwater darkness. Sonar systems used to detect submarines and other objects under water were copied from these animal abilities. So was ultrasonography.

When ultrasonic waves are bounced off body structures, the echoes vary depending on the kind of tissues they have passed through. The patterns of the echoes provide information that can be used to form a picture. It is a picture of a cross section, rather like the picture a CAT scanner produces. But doctors think that ultrasonography is probably a little safer than CAT scanning, because only sound waves are being sent through the body, not X-rays. That is why they usually use ultrasonography rather than a CAT scan on pregnant women.

Obstetricians find this tool very helpful. They use ultrasonography to get a picture of the fetus inside its mother's womb. They can tell if a woman is going to have twins or triplets. Ultrasonography also shows up certain kinds of defects in the development of the fetus, such as congenital heart defects. Then the doctors can be prepared to treat the baby as soon as it is born. Some conditions can actually be treated while the fetus is still in the womb. In hydrocephalus, for example, extra fluid builds up inside the skull and can press on the brain tissues, preventing the brain from developing properly. Recently, doctors have begun to treat hydrocephalus before birth by draining the excess fluid, using ultrasonography to show them the exact position of the fetus and to guide their needles into its head.

Echocardiography is a type of ultrasonography used to watch the movements of the heart walls and valves. Sometimes a single narrow beam of ultrasound waves is sent through the heart. The echoes of the beam are recorded in the form of a wavy line on a screen or a paper strip. Or a fan-shaped beam of sound waves may be used to produce a picture of a whole slice of the heart. (The echocardiogram, like the CAT scan, is a flat two-dimensional picture; the newer DSR scanner produces a three-dimensional image.) Heart defects can be discovered this way without subjecting the patient to catheterization, in which a hollow tube is threaded through a vein in the arm or neck into a chamber of the heart. (Unlike echocardiography, catheterization is expensive, painful, and sometimes dangerous.)

Ultrasonography can also be used to observe the size and position of body organs, to find tumors, and to spot clogged arteries. Ultrasound pictures of the eyes can pick up a detached retina and reveal tumors and foreign objects.

One interesting area of biotechnology doesn't involve sending X-rays or sound waves or any other form of energy into the body. Instead, it picks up and records the tiny bursts of the body's own natural electricity. It seems a little surprising to think that your own body is sending out electricity right now. Waves of electricity are being produced as your heart contracts, as your arms and legs move, and even in your thinking brain. These pulses of electricity can be picked up with metal wire electrodes and carried to a

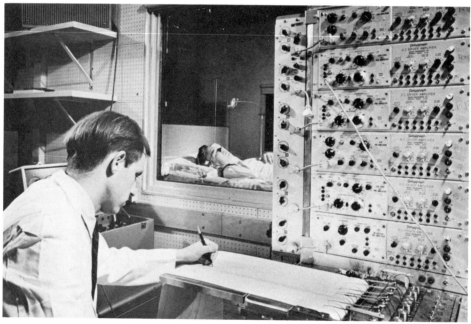

U.S. Public Health Service

While a volunteer sleeps peacefully in the laboratory, miles of paper tape record the sleeper's brain waves.

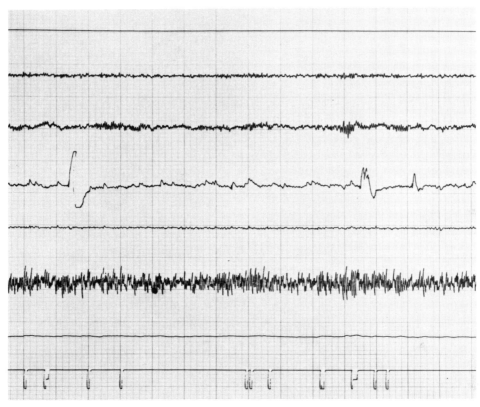

Dr. Stanley Krippner, Maimonides Hospital, Brooklyn, NY

An EEG tracing shows brain waves recorded from various parts of the head.

machine that records them in the form of wavy lines on a screen or a strip of paper. The machine that makes a record of the heart's electricity is called an electrocardiograph. An electroencephalograph records the electric "brain waves." Sometimes electrodes are inserted directly into nerve or muscle tissue. (Researchers can even record the electric pulses from a single microscopic nerve cell.) But most of the time that is not necessary. The electrodes end in small flat disks that are pasted to the skin with a jellylike substance. The characteristic zigzag waves on an electrocardiogram (EKG) can tell a heart specialist if a person has had a heart attack, or if there is some other kind of damage to the heart muscle or valves. Various kinds of brain waves

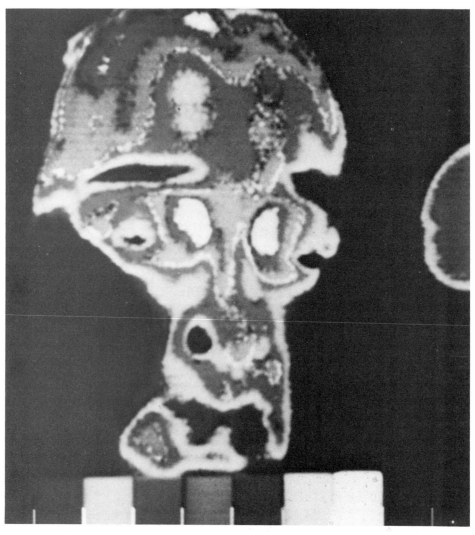

Peter Drummond and James W. Lance, Prince Henry Hospital, The University of New South Wales, Australia

Cameras with infrared-sensitive film and special color filters take thermograms of the body surface. This thermogram shows the head of a migraine headache patient; it was taken while the headache was raging. The right side of the forehead, where the pain was sharpest, shows up as the coolest area.

are recorded on an electroencephalogram (EEG), depending on whether the person is awake or asleep, thinking quietly or excited about something, or moving about actively. The electrical storm of an epileptic seizure can be traced on an EEG, and so can the calmer course of a night's sleep. A glance at the paper tape unreeling from the EEG machine can tell a sleep researcher whether a sleeper is deeply asleep or about to awake or in the midst of a dream. Electroencephalograms can also help to discover and locate brain damage from a stroke, a head injury, or a brain tumor.

Heat pictures are providing new information about the body in a technique called thermography. The thermographic camera does not work with light rays as an ordinary camera does. Instead, it records images produced by infrared detectors, which pick up the heat radiating from the body. Color filters in the camera pop into place to indicate different heat intensities, and the resulting picture is recorded on color film. Thermographic pictures highlight unusual temperature changes on the skin, which can provide clues to what is going on inside the body. Hot spots may indicate a tumor, because tumors usually have a very active blood supply, and the flowing blood carries heat. Cold spots may be a sign of dead tissue or decreased blood circulation, for example, when a blood vessel is clogged. Using thermography, doctors can detect tumors and the warning signs of strokes, or explore areas of inflammation in arthritis. They can judge how severe burns are, spot young children who have a tendency to get fat, and determine whether a person complaining of severe pain is faking or not. (Areas of pain show up as cool spots.) All this can be done without even touching the patient—just by taking pictures.

Fiber-optic probes provide a way for doctors to literally look into the body without a major surgical operation. Light travels through long flexible glass fibers like the fiberglass used to make curtains. A fiber-optic endoscope is a long flexible tube, containing two concentric cylinders of glass fiber. The outer ring of fiber transmits light down into the body, while the inner core reflects the light back to the doctor's eye or to a camera or video monitor that picks up an image of the inside of the body. A long thin fiber-optic endoscope can be threaded down the air pipes into the lungs, permitting the doctor to see signs of lung cancer and other damage. Such probes can also be used to peer up into the bladder or intestines or down into the stomach, giving a close-up view of an ulcer or tumor. By making just a tiny cut, a doctor can insert a fiber-optic probe into the body to look at organs such as the liver or pancreas. With special attachments on the tip, the probe can even be used to perform operations without opening up the whole chest or abdomen. Sterilization operations are usually performed with a fiber-optic device called a laparoscope. The doctors who started the life of the first "test-tube baby," who made headlines a few years ago, used a laparoscope to obtain eggs from her mother's ovaries. A fiber-optic fetalscope has been used to give blood transfusions to a fetus in the womb when tests showed it was suffering from a dangerous Rh reaction.

Biotechnology has given today's surgeons exciting new powers. Peering through a powerful microscope and working with delicate surgical tools, a surgeon can operate on structures literally too small to see. Doctors can now repair the tiny heart of a newborn baby. Recently, they have even begun operating on children before birth, opening up blocked blood vessels that threaten to damage the brain or kidneys. When a person's arm or leg has been cut off in an accident, delicate microsurgery can reattach it; surgeons use high magnification to help them carefully stitch together torn blood vessels and trace and rejoin fragile nerves.

Biotechnology has also provided the modern surgeon with a number of devices for "bloodless surgery." Carefully controlled ultracold temperatures can be used to freeze a cancerous tumor, which can be lifted out as a solid mass. A recently developed ultrasonic device cuts easily through soft tissues while leaving blood vessels undamaged. Laser beams, extremely narrow beams of coherent light, can be used for especially delicate operations. A laser beam can spot-weld a detached retina back into place, drill a decayed tooth, or destroy cancer cells without harming the normal tissue around them. Biological researchers are using lasers to stimulate nerve cells, to destroy particular structures inside living cells, and to snip off pieces of chromosomes—the tiny structures that carry the blueprints for heredity.

Science has come a long way since the invention of X-ray photography. Doctors can look at—and into—their patients with a variety of scanners and other devices to find out what is wrong, and they can use powerful new tools to heal and repair. Researchers are using the same tools to discover more about the healthy body and how it works. Each year brings new inventions and improvements of older ones. The tools of biotechnology are constantly providing more information in new and safer ways.

SPARE PARTS FROM "BODY SHOP"

Early in 1981 researchers at the University of Utah asked the Food and Drug Administration for permission to use an artificial heart on human patients. They had been working on their invention for years, and they had tested it on many animals. These researchers were sure their artificial heart could take over the work of a real heart, an amazing muscle that pumps steadily on throughout each person's life, keeping blood flowing through the blood vessels of the body.

In July, while the Utah researchers were still waiting for permission from the FDA, a different medical team suddenly made headlines. In Texas, heart surgeon Denton Cooley was operating on a man from the Netherlands when the patient's heart failed. The doctors couldn't start it up again. Only

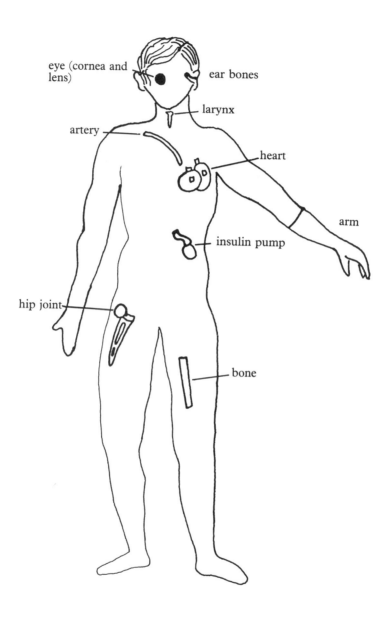

Artificial organs and body parts are helping to extend life and restore lost abilities.

Division of Artificial Organs, University of Utah, Salt Lake City

Robert Jarvik with the artificial heart he is developing at the University of Utah. (This is the Jarvik-7 model.)

a new heart could save the man. A new heart usually comes from someone who has just died in an accident. But there weren't any hearts like that available just then. Instead, Dr. Cooley decided to use an artificial heart, built by a member of his research team. The surgeons took out the patient's own heart and put the artificial heart in its place. It stayed there, beating steadily 90 times a minute, for more than two days. By then the doctors had found an accident victim whose heart could be used for a transplant. They took

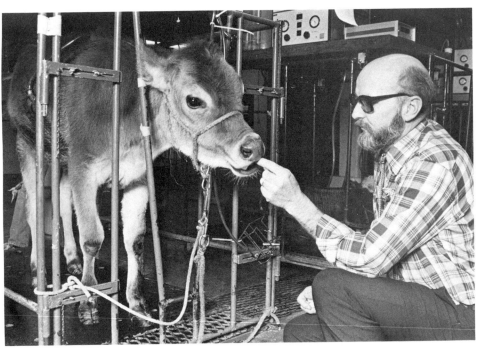

Tennyson, a calf at the University of Utah, lived for 268 days with an artificial heart. (The compressed air lines that drive the pump can be seen on his back.)

out the artificial heart, which was still working perfectly, and carefully attached the heart from the accident victim. The patient lived for another week, but then he died of an infection.

Neither Dr. Cooley nor the Utah team headed by Willem Kolff intended to use an artificial heart permanently. They just wanted a heart that could take over in an emergency and keep the patient alive until a transplant heart was available. But some day soon, they hope, there will be better artificial hearts that will continue to work for the rest of a person's life.

Many problems had to be solved in developing the artificial hearts researchers are using today. The heart is an amazing machine that works on tirelessly, pumping about 2,000 gallons of blood each day. An artificial heart has to be just as tough and long-wearing, but it must also be gentle enough not to harm the delicate red blood cells that carry oxygen to the body.

28

It must be made of materials that will not irritate the body tissues or cause infections. Most of these problems have been solved in the newest artificial hearts, especially the one designed by Robert Jarvik, a member of Kolff's team in Utah. Calves there have lived for as long as nine months with artificial hearts beating inside their bodies. But one big problem still remains. These calves' artificial hearts are driven by compressed air. A patient with such a heart would have to be attached—all the time—by wires and hoses to big machines and compressed air tanks. Now Jarvik is working on a better model, with a motor and pump about the size of a flashlight battery, powered by a battery pack that the patient can wear on a belt around the waist. Special electronic sensors will be built in, so that the artificial heart can pump faster or slower to meet the body's needs.

Full artificial hearts are not yet ready for wide use, but artificial heart *parts* have been saving people's lives for many years. Sometimes after a heart operation a person's heart is not strong enough to take over. If only it had a chance to rest, perhaps for a few days, it could recover. But a living heart has no chance to rest. If it stops pumping, blood stops flowing through the body. Then body cells begin to starve. They need the food materials and the oxygen the blood brings them. The body cells also need the flowing blood to take away their waste products, which could accumulate and poison them. Within minutes after the heart stops, cells begin to die, especially the nerve cells in the brain. If too many cells die, the person will die.

During heart surgery, doctors can stop the heart without killing the patient by sending the blood through tubes leading to and from a heart-lung machine. This machine pumps the blood from the body and exchanges the waste carbon dioxide it carries for fresh oxygen that the body needs. But a person cannot stay hooked up to a heart-lung machine indefinitely, because it can damage the delicate red blood cells. Now medical researchers have developed a special pump that can be attached temporarily to take over the work of an ailing heart for a few days. This device is called a left ventricular assist pump. (The ventricles are the thick-walled lower parts of the heart, which do the main work of pumping blood. The left ventricle pumps blood through the main blood vessels of the body, while the right ventricle sends blood from the heart to the lungs to exchange carbon dioxide and other waste

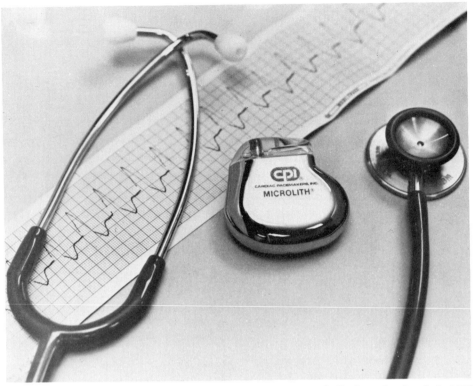

This artificial pacemaker, shown next to a stethoscope and EKG tracing, can substitute for a faulty natural pacemaker and keep the heart beating in a steady rhythm.

materials for oxygen.) A left ventricular assist pump is not meant to replace a heart, but only to help it out until it is ready to take over again.

The steady rhythm of a normal heartbeat is set by a built-in timing mechanism called a pacemaker. The body's natural pacemaker is a specialized group of cells in the heart wall that send out tiny pulses of electricity—not enough to give a person a shock, but strong enough to act as a signal, telling the heart muscle to contract. Sometimes the natural pacemaker does not work as well as it should, and the heart misses beats or fails to keep to a regular rhythm. A heart with a faulty pacemaker does not pump blood as efficiently as it should, and it may not pump enough blood for the body's needs. Since the late 1950s, doctors have been able to help such patients by inserting artificial pacemakers. More than a hundred thousand people now have artificial pacemakers, and over the years these devices have been constantly improved. Today's pacemakers can stay in the body for many years without either hurting the body tissues or being damaged by the body fluids. Long-lasting batteries have been developed, so that a person no longer needs another operation to get the pacemaker battery changed. Some batteries can be recharged from outside the body. Some pacemakers are programmable: their rate can be adjusted by radio signals. For example, the heart rate could be speeded up to meet the body's increased needs during running. (The older pacemakers could set only one steady rhythm.)

Another key part of a working heart is the set of valves that control the flow of blood from one chamber of the heart to another or to the large arteries that carry blood away. When heart valves are damaged, perhaps by a disease such as rheumatic fever, today's surgeons can replace them with plastic valves, or sometimes with specially treated valves from pig hearts. Surgeons can also stitch in flexible tubes of artificial materials to replace damaged parts of arteries.

Heart transplants and artificial hearts made headlines because the heart is such an important organ. Actually, though, researchers have been making progress on many other artificial organs and body parts. Willem Kolff, who is at the forefront of work on the artificial heart today, was a pioneer in developing an artificial kidney. He built the first dialysis machine back in the 1940s, when he was a young doctor in the Netherlands.

The kidneys are two bean-shaped organs that remove nitrogen compounds and various other waste products and poisons from the blood that flows through them. These waste products then pass out of the body in the urine. The kidneys filter about 180 quarts of liquid each day. Without their work, poisons would build up in the body. A person whose kidneys have been badly damaged by disease or accident is doomed to die unless there is some other way of filtering out the body's wastes and poisons. The dialysis machine that Willem Kolff built does just that. The person's blood passes out through tubes to a machine where it is pumped through a large artificial

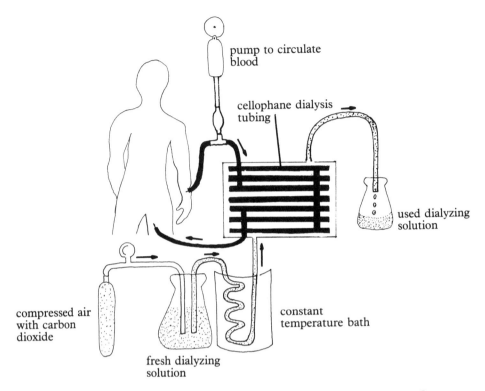

pump to circulate blood

cellophane dialysis tubing

used dialyzing solution

compressed air with carbon dioxide

fresh dialyzing solution

constant temperature bath

In kidney dialysis, a collection of tubes and filters cleans wastes and poisons from the blood, substituting for the natural kidneys.

membrane that holds back wastes and poisons. Then the purified blood passes back into the person's body.

The first artificial kidney machines were very crude. In fact, the membranes were pieced together from cellophane sausage casings. In the years since then, the design of dialysis machines has been improved many times. The huge machines of the past can now be replaced by much smaller units, some even small enough to be worn as backpacks. But patients who need an artificial kidney still must spend many hours a week hooked up to a machine. They can read or listen to music or do other quiet things while their blood is being cleansed, but they cannot live a fully normal life. Researchers are working on various designs for a real artificial kidney, one that can be implanted inside the body. So far they have been having trouble fitting enough filtering membrane into such a small space. One promising new approach is to have the person swallow tiny granules of specially treated charcoal covered with a plastic material. As these tiny plastic "pills" travel through the stomach and intestines, wastes and poisons are trapped in the charcoal and pass out of the body.

Another important organ in the body is the pancreas. It produces chemical messengers called hormones that help to control how the body uses sugar. Sugar is the body's main energy food. Sugars in the foods we eat may be combined with hormones ("burned") to give energy, or they may be stored away for future needs in the form of a starch. Insulin, one of the hormones of the pancreas, is released when there is a lot of sugar in the blood— for example, just after a meal. Insulin helps the body cells to take in sugar, and it also signals the liver to store more sugar. So this hormone makes the blood sugar level go down. People with diabetes do not produce enough insulin, or the insulin they make does not work properly. Their bodies cannot use sugar effectively, and they may become seriously ill or even die. Many diabetics are helped to live a fairly normal life by taking daily injections of insulin.

There is a problem with insulin injections, though. They shoot a large amount of the hormone into the body all at once. This is not the way the pancreas works. The normal organ lets out its hormone in tiny amounts, whenever signals of a high blood sugar level tell it that more insulin is needed. Diabetics can smooth out their insulin and sugar levels somewhat by taking several injections a day, using mixtures of various kinds of insulin that work for different lengths of time and timing their injections so that the sugar surges after meals are covered by larger amounts of insulin. Even so, the insulin supply does not always perfectly match the body's needs, and taking several injections every day is a troublesome way to live. A number of teams of medical researchers are working on designing an artificial pancreas that will work more like a real pancreas.

Table-top-size models, such as the Biostator, have already been built. They include a sensor to measure the person's blood sugar level, an insulin reservoir, a pump to shoot measured doses of insulin into the body through

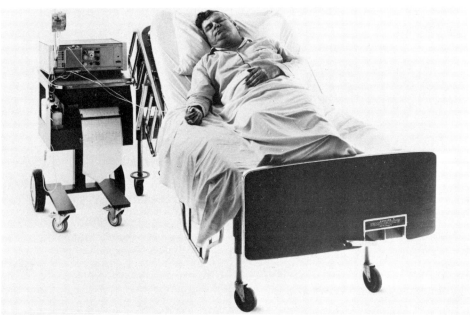

Life Science Instruments, Miles Laboratories, Inc., Elkhart, IN

The Biostator is a complete artificial pancreas system, which monitors the amount of sugar in the blood and injects the right doses of insulin. But it is too large to carry around.

a needle that stays permanently inserted in a vein, and a tiny computer to figure out exactly how much insulin the body needs at any particular time. Big machines like these are only good for bedside use in hospitals. For daily living, researchers are trying to pack the same kind of parts into a much smaller package. Belt-pack models have now been built, but the researchers' goal is to make an artificial pancreas small enough to be implanted in the body. An implantable insulin pump, with a reservoir that can be refilled from outside the body, is already being used to release insulin slowly and steadily. But the diabetic must still measure blood and urine sugar levels to set the dose. The next step will be an artificial pancreas with its own built-in sugar sensor and computer to do the job automatically.

A few years ago, pollsters asked people what medical condition they feared most. The largest number of people answered ''cancer.'' But the sec-

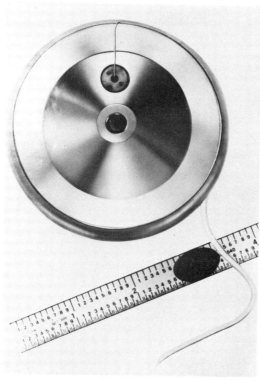

The implantable insulin pump delivers tiny measured doses of insulin that help keep the blood sugar steady. But it does not have a built-in sugar sensor.

ond most frequent reply was "blindness." In another study, volunteers who wore special earplugs that made them hard of hearing became fearful and suspicious of other people. Our eyes and ears are very important organs. People can live without them, but their lives are greatly limited. In recent years, technology has brought exciting new approaches to the treatment of blindness and deafness.

In the living eye, light rays pass through a clear outer covering called the cornea and are focused through a structure called the lens onto a sensitive layer at the back of the eyeball, the retina. There the light rays stimulate special nerve endings that send messages along the optic nerve to the visual

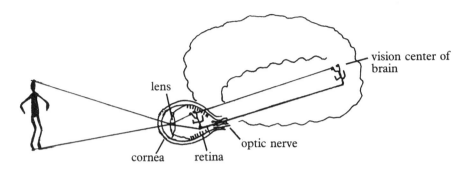

The human eye: the lens focuses and transmits light rays to the sensitive retina at the back of the eyeball. Nerve cells in the retina relay messages about the image to the vision center at the back of the brain, where they are translated into a meaningful picture.

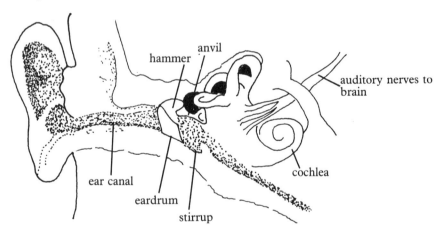

The human ear: vibrations of the eardrum, produced by sound waves, are transmitted by three little ear bones to a fluid-filled chamber containing the cochlea. There the vibrations stimulate sensitive cells that send messages to the hearing center in the brain.

center at the back of the brain. The brain sorts out all these nerve impulses into meaningful pictures.

In the ear, sound waves set up vibrations in the eardrum, the thin membrane that covers the entrance to the middle ear chamber. The vibrations are passed on first to three little ear bones and then to the fluid that fills an inner ear chamber called the cochlea, which is coiled like a snail shell. Sensitive nerve cells in the lining of the cochlea respond to particular sound frequencies and send messages to the hearing center in the brain. There the messages are sorted out into speech, music, and the other sounds that fill our world.

Accident or disease can damage various links in the chain of structures by which we hear and see. Artificial replacement parts can bring sight again to many of the blind and hearing to the deaf. If a cornea is badly scarred, for example, it can be replaced with an artificial cornea of clear acrylic plastic, a material similar to that used for the windshields of jet planes. An eye lens that is clouded by cataract may be replaced by an implantable plastic lens. The artificial lens cannot be refocused by the eye muscles for near or far vision as the natural lens can, but when it is combined with eyeglasses, the person can see quite well.

Plastic eye parts cannot restore vision when the retina or the optic nerve is damaged. But several groups of researchers are working on "bionic" vision systems that may eventually be real substitutes for the eyes. A team at Smith-Kettlewell Institute of Visual Sciences in San Francisco is developing a Tactile Vision Substitution System (TVSS) that lets people "see" with their skin. A miniature TV camera mounted on a pair of eyeglasses transmits a picture to a vest on which more than a thousand small cones are mounted. The cones vibrate, producing a pattern of dots that the person feels on the skin of the belly. It takes some practice to be able to translate these skin sensations into mental pictures. But blind students have learned to "see" with the TVSS well enough to find objects around a room, read meters, and even use scientific instruments.

William Dobell at the Institute for Artificial Organs in New York is using another approach, closer to the way people normally see. He implants electrodes in the visual center of the brain. The person's visual field is then mapped to find out which electrodes correspond to points that are "up" or

A blind volunteer is learning to use an artificial eye system at the University of Utah. The X-ray on the next page shows how the electrodes directly stimulate the vision center at the back of the volunteer's brain.

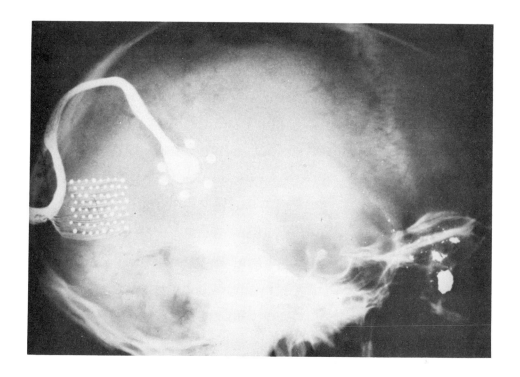

"down," in the middle or to the sides. (Each person has his or her own individual vision map.) After that, a TV camera transmits pictures that are translated by a computer into points on the "map"; the correct electrodes are stimulated, and the person "sees" images. Blind people using the device have been able to read Braille characters five times as fast as they can with their fingertips, but the images are still too crude to permit reading ordinary print. In addition, the computer and TV camera are too large to carry around. But eventually the researchers hope to develop a model with a camera small enough to fit inside a glass eye and a computer that can be mounted on an eyeglass frame.

Meanwhile, biomedical engineers have developed many devices to aid blind people. An Optacon scans pages of print or computer displays and converts the letters, through a computer, into vibrations that the user can feel through the fingertips. Other devices convert printed material to spoken words. A variety of canes and other devices help blind people get

The Sonicguide, a device that fits into an eyeglass frame, gives the blind user electronic signals of obstacles in the way.

around by bouncing beams of light or ultrasound off objects; the reflections are converted to sounds or vibrations that warn of obstacles in the way.

Electrical hearing aids that amplify sounds date back to the beginning of the century. Today's aids for the deaf go much further. Some forms of deafness result from the three little bones in the middle ear chamber growing together so that they cannot vibrate. People with overgrown or damaged ear bones can be helped to hear again with Teflon or bioglass replacements. In other forms of deafness, a whole new sound-transmitting system may be needed. Research teams are working on a bionic ear. It consists of a microphone implanted in the ear or skull, a miniature computer that breaks up the electrical signal of the microphone into different frequencies, and tiny electrodes implanted in the sensitive lining of the cochlea. Using one model, developed in Australia, deaf people can hear a telephone ringing and even make out the words of speech.

A person whose larynx (voice-box) has been removed because of cancer can now speak clearly with an artificial larynx developed by medical researchers in Philadelphia. The device, which produces sounds electronically and sends them through a miniature amplifier and speaker, fits into a dental plate in the roof of the mouth. Complete with batteries, it is no bigger than a half dollar. Hand-held voice synthesizers are also available. The user presses keys to produce various combinations of spoken syllables, words, and common phrases. People who are paralyzed or handicapped by cerebral palsy can communicate using other modern electronic marvels: typewriters and computer terminals that can be operated by the breath, the touch of a finger or a tongue, or even by the blink of an eye.

From time to time you'll see newspaper stories about surgeons who work for hours to reattach a person's hand or foot that was cut off in an accident. The tools and techniques of the delicate microsurgery that make this kind of dramatic feat possible are fruits of the biotechnology revolution. But often a lost arm or leg cannot be saved, and an artificial replacement is needed. Bioengineering researchers have come a long way from the wooden ''peg leg'' and the crude metal hook that are familiar from scary tales of pirates.

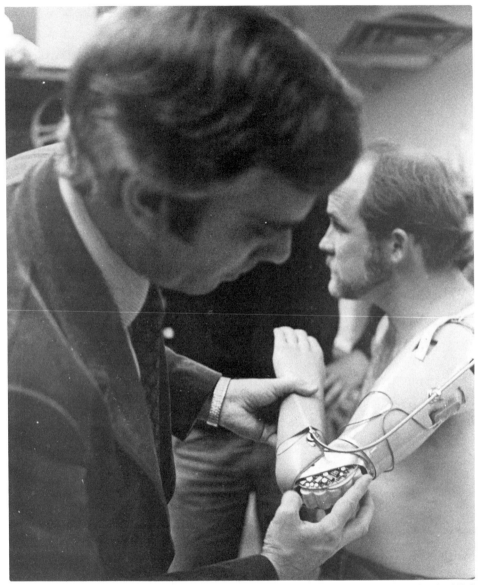

The new artificial arms, such as the "Utah arm" shown here, provide feedback sense messages and use impulses from the nerves of the stump to control the arm's movements.

42

Special metal alloys and space-age plastics are used to fashion replacement bones and joints, and the newest artificial limbs are going bionic.

Several teams of researchers, for example, are developing "myoelectric" arms. The artificial arm, covered with realistic-looking and natural-feeling skin-colored plastic, contains a tiny electric motor, a rechargeable battery, and a miniature computer housed inside a control box the size of a deck of cards. When you move your arm, tiny electrical impulses carried along nerves that run through your shoulder tell some of your arm muscles to contract. Nerve messages from your skin and joints travel back up your arm and tell you exactly where the various parts of your arm are at each moment. The myoelectric arm works the same way. Sensors attached to the stump pick up the tiny electrical impulses from muscles and nerves and transmit them down to the computer, which translates them into movements. Sensors in the arm report back to the nerves in the stump with information on position, so that the movements can be adjusted just like the movements of a natural arm.

Computerized artificial legs that work in a similar way are also being developed. The movements of the artificial leg can be coordinated with sense messages from the other normal leg, so that the person will walk smoothly. Leg braces designed in this way allow people who are paralyzed to walk.

Biomedical engineers are also working on artificial replacements for skin, blood, organs such as the liver and lungs—just about anything in the body that may need replacing. What about the brain, the most important organ in the body? Will we ever have artificial replacements for the brain?

The soft wrinkled mass of nerve cells nestled inside your skull contains all your thoughts and memories, the experiences and personality patterns that make you uniquely *you*. Today's researchers are far from being able to duplicate anything as complicated as a human brain. In fact, they don't all agree on exactly how memories are stored—in the form of electrical patterns or chemical "tags" or, more likely, a combination of both. But brain researchers have learned a great deal. For example, by observing what happens when particular parts of the brain are damaged or destroyed and by stimulating certain parts of the brain with chemicals or electrodes, they have drawn up

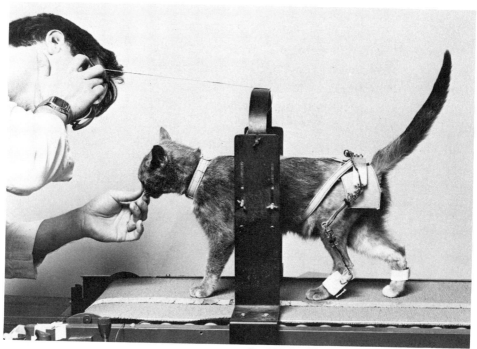

Jerrold S. Petrofsky, Wright State University, Dayton, Ohio

This cat, wearing a leg brace, is walking on a treadmill, while electrical impulses transmitted to a computer provide a record of its movements. Researcher Jerrold Petrofsky is analyzing the results of studies like these to devise a computerized system that will help paralyzed people walk.

"brain maps." They know which parts of the brain control the movements of various body parts and which parts receive and sort out the various sense impressions. They have found areas of the brain that control speech, arithmetic ability, musical ability, and particular kinds of memory. Various kinds of brain pacemakers are being developed, sending out electrical impulses that can take over for a damaged control area in the brain. In a condition such as cerebral palsy, an electronic pacemaker can block faulty messages from the brain that make the muscles contract spastically, and thus help the person to move and walk more normally.

Studies of "brain waves," the weak pulses of electricity that can be picked up with electrodes pasted to the surface of the head, have yielded a great deal of interesting information and some useful applications. If people are told when they are producing a particular pattern of brain waves, such as the alpha waves that are recorded when one is quietly alert, they can learn to produce the same patterns again whenever they want to. This technique, called biofeedback, has been used to help people learn to relax and to relieve headaches. With biofeedback, people can even do things like lowering blood pressure and decreasing the amount of acid produced in the stomach. If the electrodes are connected to electrical switches set to work when a particular brain wave pattern is produced, people can learn to turn light switches or machines on and off just by thinking at them. (This trick may be useful in designing ''smart wheelchairs'' and other devices for use by handicapped people.)

Some groups of researchers have been studying patterns of brain waves to try to identify particular thoughts. They are still far from being able to read people's minds, but they have been able to pick out typical patterns for certain kinds of thoughts. For example, there is a ''surprise wave'' when something unexpected happens, like the sudden ringing of a doorbell. A different type of brain wave appears when a person is concentrating on one voice in a crowd of people talking. (When people drink alcohol, that particular brain wave gets smaller—they can't concentrate as well.) This is the kind of information researchers will need if they ever hope to build a replacement brain.

Another way of learning about how the brain works is to build artificial brains. The computers in use today have electronic ''brains.'' Most of them are just high-powered adding machines. They can do certain tasks, such as mathematical calculations, much faster than humans can, but how well they work depends on their ''program''—a plan of action designed by humans. These computers do not think for themselves. However, some experimental computers have been designed to solve problems and learn new information in much the same way we do. Other computers have been programmed to carry on conversations so realistically that people talking to them by typewriter cannot tell which answers were given by a computer and which by another human. Computers have been taught to play games like checkers and chess, to draw pictures, and even to write music and poetry. In the automobile industry and some others, robots with simple computer brains sort and assemble parts and do jobs like precision cutting and welding. Researchers have built experimental robots that can learn to obey simple commands, move about, and fetch objects. So far, though, the electronic brains that have been built have not been as complex as the human brain.

TINKERING WITH HEREDITY

In July 1974 a group of scientists working in a field called recombinant DNA research did an unusual thing. They asked their colleagues to stop doing certain kinds of experiments because they thought they might be dangerous. They called for a meeting of experts in the field to discuss the possible dangers and decide what to do about them.

All over the world, researchers stopped their work and prepared for the meeting. It was held at Asilomar, in California, in February 1975. Scientists talked and argued for three days and nights. They finally decided that they didn't know whether or not certain kinds of experiments were safe. They said that some experiments should not be conducted at all until more could be learned about possible risks. Other experiments should be done only in

Recombinant DNA research (here and on the facing page) requires different degrees of safety precautions, depending on the possible risks.

special laboratories with air-filtering systems and safety cabinets where researchers wore gloves when they handled the experimental materials.

A committee was set up by the National Institutes of Health to draw up guidelines for the new recombinant DNA work. Meanwhile, angry debate continued. Some scientists feared that the experiments might produce terrible new diseases and other unknown dangers. Others said those fears were silly, and all the restrictions were wasting precious time that could be used for experiments that would bring new cures for diseases and many useful industrial products.

The argument made headlines again in 1976, when Mayor Alfred Vellucci of Cambridge, Massachusetts, found out that Harvard University planned to convert a laboratory for recombinant DNA work. Mayor Vellucci called a public hearing, and the city council voted to ban recombinant DNA

work temporarily while a committee of doctors and laymen heard testimony from the experts and decided whether this research would be a health hazard for the people of Cambridge.

However, in the years that followed, people decided that recombinant DNA was safe, and it is now considered one of the truly great advances in science.

What is recombinant DNA? Why was it thought to be dangerous? What benefits could the experiments bring?

To understand the problems and the promise of recombinant DNA research, it is necessary to know some basic facts about the body, the microscopic cells that compose it, and the way traits are passed on by heredity.

Each human being starts life as a single cell. This tiny cell divides to

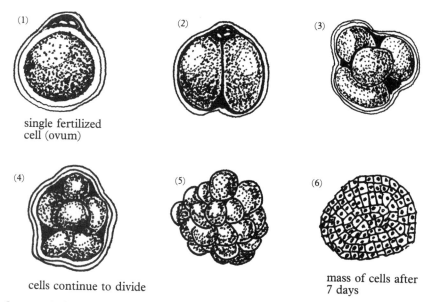

(1) single fertilized cell (ovum)

(2)

(3)

(4)

(5)

(6)

cells continue to divide

mass of cells after 7 days

A human being starts off as a single cell, which divides and grows, producing a mass of cells that burrows into the lining of the mother's uterus.

form two cells, which divide again and again. A solid ball of cells forms, and then a hollow ball. As the cells continue to grow and divide, some of them begin to change. Gradually a tiny creature takes shape, looking more and more human as it grows. Your body contains trillions of cells. The bodies of animals and plants also contain many cells, but there are other creatures in the world that are made up of only a single cell. Bacteria and yeasts are one-celled organisms.

What determines whether a particular cell will be a bacterium or whether it will grow into a dog or a corn plant or a human being? Each cell contains a complete set of instructions for the kind of organism it belongs to. These instructions are spelled out in a chemical code, and they cover not only how the creature will develop, but also all the chemical reactions that go on in the cell. Before a cell divides to form two cells, it carefully copies all of its hereditary instructions. Then, when the cell divides, each of the new cells receives a complete set of instructions of its own. Every one of the trillions of cells in your body has a copy of the complete set of instructions for mak-

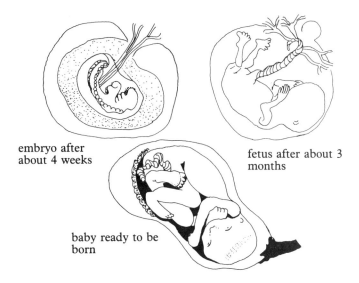

embryo after
about 4 weeks

fetus after about 3
months

baby ready to be
born

The embryo grows and develops into a fetus, and finally a baby.

ing you. But you have many different kinds of cells—long, slim nerve cells, flat skin cells, red blood cells that look like tiny doughnuts without the hole, and many others. How could they be so different when they all have the same instructions? The answer is that in any cell most of the instructions are "turned off," and only the ones for that particular cell are working.

The cell's hereditary instructions are coded in a chemical called deoxyribonucleic acid, or DNA. A DNA molecule is a long, complicated double chain. Each link in each strand of the chain is called a nucleotide. There are four main kinds of nucleotides. So the DNA code is spelled out in an "alphabet" of four letters.

It seems strange to think that an alphabet of only four letters could spell out enough different messages to make a hamster, an apple tree, and all the other forms of life. But there are many different ways to arrange even four letters. A whole DNA molecule may have hundreds of thousands of links in its chain. It is hard to imagine how many ways that number of letters could be arranged, even if you are working with just four kinds of letters.

Most of the DNA in a cell is found in structures called chromosomes. Human cells have 46 chromosomes. Under a microscope, they look like tiny rods. A single-celled bacterium has only one chromosome. The ends of its DNA chain are joined together, so that the bacterial chromosome looks like a ring.

For the cell to do its work, the DNA instructions must be translated into a different kind of chemical: protein. Some proteins form the structures of cells, and those of the body. Hair, for example, is almost pure protein. Some proteins, called enzymes, make the cell's chemical reactions work. Some proteins, called hormones, carry messages from one cell to another and help to keep the body running smoothly.

Proteins are made at tiny structures inside the cell called ribosomes. To make each protein, a particular portion of the DNA instructions is carried to the ribosomes by a chemical messenger called RNA (ribonucleic acid). As you might guess from its name, RNA is rather similar, chemically, to DNA. It, too, is made up of a long chain of nucleotides, but it is a single chain.

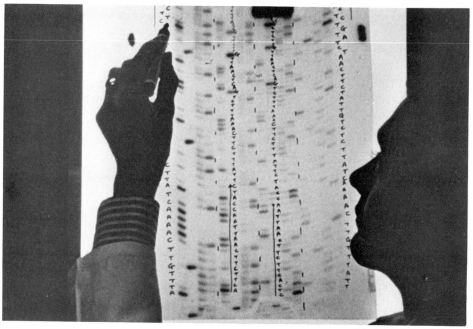

G.D. Searle & Co., Skokie, IL

*Today's researchers can take apart **DNA** molecules and map the genes, reading out the coded sequences of nucleotides.*

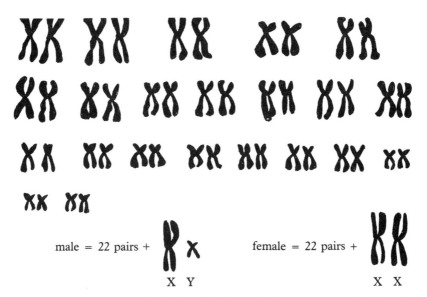

male = 22 pairs + XY

female = 22 pairs + XX

The human chromosome set includes 23 pairs of chromosomes. Each cell in the body contains a complete set of 46 chromosomes. (These chromosomes have been sorted and arranged according to shape and size; in a real cell they are randomly mixed.)

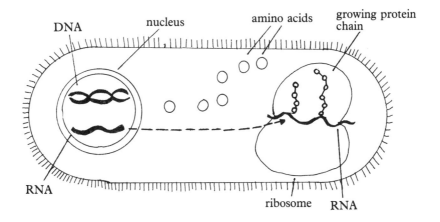

DNA carries the cell's hereditary instructions. RNA molecules are made according to the DNA blueprints. Then, at the ribosomes, proteins are constructed from amino acid building blocks according to the instructions coded in RNA.

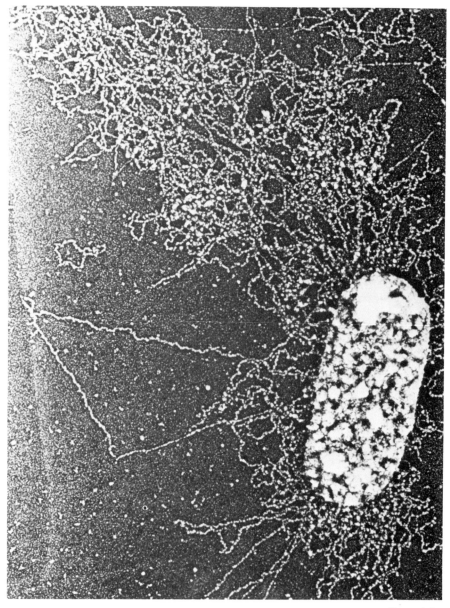

A highly manified view of E. coli, *a microscopic bacterium that can live in human intestines; it is widely used in laboratory studies and recombinant DNA syntheses.*

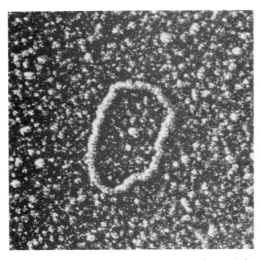

A plasmid, shown highly magnified, is a circular DNA molecule used in gene splicing.

Proteins are also made up of long chains. They are formed from building blocks called amino acids. There are about 20 kinds of amino acids in proteins. They could spell out an enormous number of different protein messages.

The portion of DNA containing the instructions for making a particular protein is called a gene. The RNA molecule that carries these instructions to the ribosomes is called messenger RNA. When the DNA message is copied onto RNA, that is called transcription. Putting together amino acids to form a protein according to the RNA blueprint is called translation.

Now we can come back to the question: What is recombinant DNA?

Recombinant DNA is a kind of "gene splicing." Genes—pieces of DNA —are inserted into the DNA of bacteria, reproduced there, and eventually translated into proteins. The genes that are inserted need not have come from bacteria. Researchers have used recombinant DNA techniques to put genes from other forms of life—toads, fruit flies, plants, mice, and even humans— into bacteria.

In 1973 two teams of researchers in California, led by Stanley Cohen at Stanford University and Herbert Boyer at the University of California at San

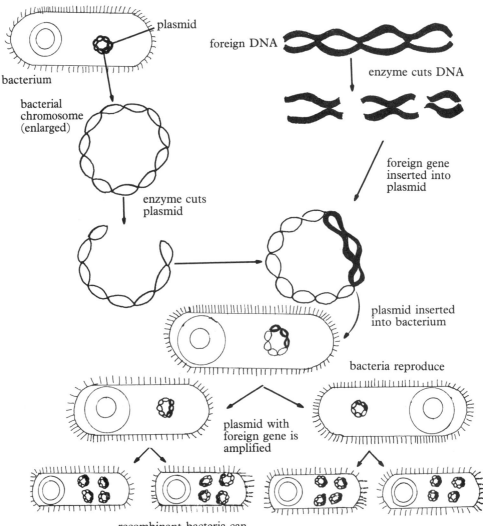

plasmid

bacterium

foreign DNA

enzyme cuts DNA

bacterial chromosome (enlarged)

foreign gene inserted into plasmid

enzyme cuts plasmid

plasmid inserted into bacterium

bacteria reproduce

plasmid with foreign gene is amplified

recombinant bacteria can produce the foreign protein

How recombinant DNA works: a plasmid from a bacterium is cut with enzymes, and then a "foreign" gene is spliced in. The "recombinant" plasmid is then put back into a bacterium, where it can be reproduced to make many copies, all containing the foreign gene. Under the right conditions, the bacterium will make the protein for which that gene carries instructions.

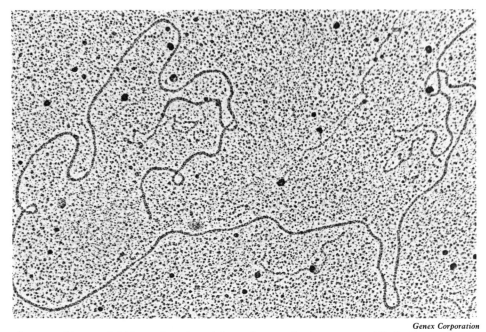

A recombinant DNA molecule with an inserted piece, magnified about 80,000 times.

Francisco, made world headlines. They were working with a bacterium called *E. coli*, which was originally taken from human intestines. *E. coli* is a favorite in laboratory studies, and a great deal is known about it. Its single cell is rod-shaped, and it contains a single large ring-shaped chromosome, which codes for a few thousand genes. In addition to this large chromosome, *E. coli* also contains smaller rings of DNA, big enough to hold a dozen genes. These small rings are called plasmids. Many bacteria have plasmids, and so do yeasts.

The California scientists used two key discoveries of earlier researchers: two kinds of enzymes. One kind, called restriction enzymes, can cut a double strand of DNA, leaving sticky ends dangling. The other kind of enzyme used in recombinant DNA work is DNA ligase, which joins broken DNA chains together when the sticky ends are matched up.

In cutting and splicing DNA, the California research teams used these two kinds of enzymes as a sort of "scissors" and "glue." They cut up DNA molecules into convenient-sized pieces with restriction enzymes. They also used these enzymes to open up plasmid rings. Then they mixed the opened-up plasmids with the pieces of DNA containing the new genes. The sticky ends matched up and were glued together with DNA ligase. In this way new plasmids were formed, containing the added genes.

A bacterium reproduces by dividing into two new cells. Before it divides, it copies its chromosomes and also copies any plasmids that it contains. Researchers have discovered ways to make bacteria produce extra copies of plasmids—even thousands of them. If the plasmid contains an inserted gene, the bacterium will make many copies of the gene. This process is called gene amplification.

When the conditions are just right, bacteria can divide very quickly—every twenty minutes or so. One original bacterium can rapidly give rise to millions. Each of them contains an exact copy of all the DNA that the original bacterium had. Scientists call a group of bacteria all descended from one original parent, and all possessing the same hereditary information, a clone. So when they use recombinant DNA techniques to make many copies of a gene inserted into a plasmid and then reproduce the bacterium, they say they are "cloning" the gene.

Cohen and Boyer and their research teams originally used the recombinant DNA technique to get large amounts of genes so that they could study them more easily. Normally, any particular gene is present in a cell in very tiny amounts, and it is mixed with thousands of other genes. The protein that the gene codes is likewise produced in very tiny amounts, and it is in a complicated mixture of chemicals that is very difficult to separate. But cloning can be used to single out just one gene and its protein product. Then the recombinant DNA technique can be used not only to study genes and proteins, but also to produce large amounts of proteins. Some of these biochemicals are valuable products for medicine, industry, and agriculture.

Where do recombinant DNA researchers get the genes that they put into bacteria? They may cut up chromosomes from some interesting cell—say, a mouse cell or a human cell—and clone the pieces of DNA. Or they may start with messenger RNA and use it as a blueprint for building DNA. (The enzyme they use for this is called reverse transcriptase. It makes transcription go backward—the reverse of the usual DNA-RNA process.) Researchers can even build genes "from scratch," starting with the individual nucleotides.

58

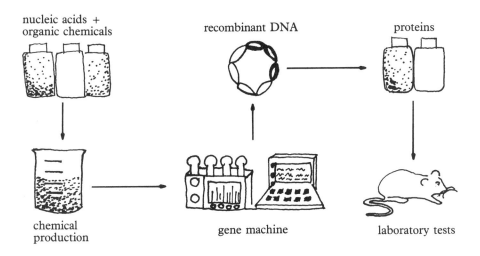

nucleic acids +
organic chemicals

recombinant DNA

proteins

chemical
production

gene machine

laboratory tests

The first gene synthesis won a Nobel prize, but today researchers can use an automatic gene machine and then translate the genes into drugs and other useful products.

If they know the way the amino acids are arranged in a protein, they can figure out the DNA message in the gene for that protein.

Working out the first amino acid sequence of a protein was a long and complicated task. It won the British biochemist Frederick Sanger a Nobel Prize. Working out the "genetic code" for translating genes into proteins involved even more years of research and also won Nobel Prizes. Today's researchers can enjoy the fruits of all that hard work. There are even machines to do the job for them. Automated protein sequences can now turn out the amino acid sequence of a protein with push-button ease within a matter of days. (It took Frederick Sanger ten years to work out the sequence of the small protein insulin.) The first synthesis, or artificial construction, of a gene was a tremendous feat. It was accomplished in 1970 by molecular biologist Har Gobind Khorana and won him a Nobel Prize. But now, for less than $30,000, a laboratory can buy an automatic "gene machine." All the researcher has to do is type out the genetic code for a particular gene on the machine's keyboard. A computer directs the automated machinery that builds up the gene, one nucleotide at a time. It takes about half an hour to add each nucleotide.

Recombinant DNA researchers have already produced a number of useful and important biochemicals through bacteria. These include human growth hormone, which can help some very short children to grow to a normal size; human insulin, a hormone that diabetics need; and thymosin and interferon, substances that play an important role in the body's defenses against disease. Before recombinant DNA techniques were developed, medical researchers had very little of these chemicals to work with. Now that large amounts of these biochemicals are becoming available, the researchers hope that thymosin and interferon, in particular, will be new wonder drugs for the treatment of infections and cancer. Thymosin may even help to slow down the aging process and keep people young and healthy.

Plasmids are not the only way to introduce genes into cells. Viruses can also pick up "foreign" genes and carry them into the cells they infect. This happens often in nature, but scientists can also introduce genes into viruses in the laboratory. A Nobel Prize was awarded recently to Paul Berg for recombinant DNA experiments using viruses to carry genes.

Viruses can carry genes not only into bacteria and other single-celled organisms, but into other cells as well. They can even carry genes into human cells. Medical researchers hope that people with genetic diseases will some day be helped by viruses that will carry into their cells new genes to substitute for faulty ones. A quarter of a million American children are born each year with birth defects, and millions of people suffer from hereditary diseases. Often just a single gene, producing a single protein, is the cause.

People with sickle-cell anemia, for example, produce a faulty kind of hemoglobin. (This is the red substance in red blood cells which carries oxygen.) Instead of a plump doughnut shape, the red blood cells of people with this disease sometimes form a sickle shape, like tiny crescent moons. Sickle cells may stick together and clog blood vessels. Sickle-cell anemia may result in crippling and early death.

Actually, the change in the hemoglobin gene in sickle-cell anemia is very small. Just one amino acid out of about 300 in the hemoglobin molecule is changed. Researchers hope that if they can put genes for normal hemoglobin into the blood-forming tissues of people with sickle-cell anemia, the disease will be cured. Their bodies will then produce enough normal hemoglobin to keep their red blood cells from "sickling."

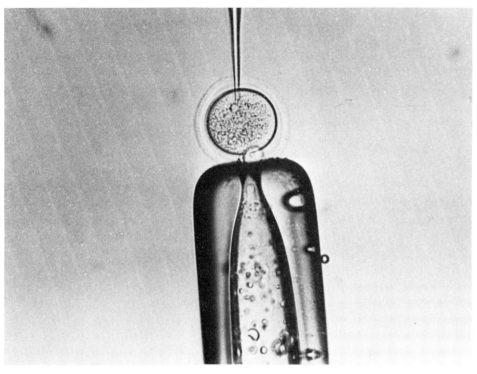

Dr. Jon W. Gordon, Department of Biology, Yale University

Using a very thin capillary tube, the researcher injects genes into a mouse egg cell. (The photo is greatly magnified.)

Viruses are one way to introduce genes into cells, but there are also other ways. Researchers have even tried injecting genes right into cells. They use a very thin glass tube for the injection. The tip of the tube is only a thousandth of a millimeter wide. The work has to be done under a microscope, of course, because not only the genes but even the cells themselves are too small to see with the naked eye. One research group, at Ohio University, injected a rabbit gene for a portion of the hemoglobin molecule into mouse egg cells. The mice that grew from the cells still had the rabbit gene, and their blood cells contained some of the rabbit protein. When they mated and had young, some of their offspring inherited the rabbit genes and made the rabbit protein. In other laboratories, human genes have been injected into mouse egg cells, and in some cases the bodies of the mice produced human proteins.

61

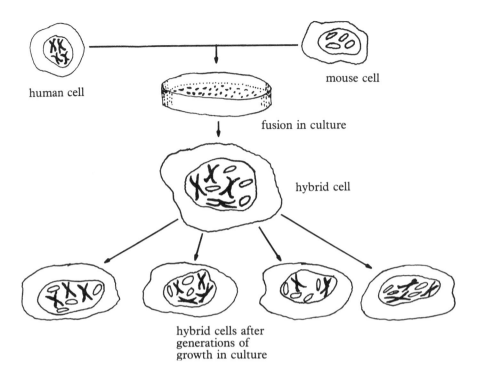

human cell

mouse cell

fusion in culture

hybrid cell

hybrid cells after
generations of
growth in culture

*Cell fusion: under special conditions, cells from two different species
can be made to join in a culture dish, forming hybrid cells carrying the
heredity of both "parents." When such hybrid cells reproduce, they tend
to lose some of the chromosomes.*

Another method researchers can use to transfer genes is called cell fusion.
Specially treated cells literally join together, merging their contents. The
hybrid cell that results contains the chromosomes of each of the original
cells that formed it. Some very unusual hybrid cells have been produced in
the laboratory, such as mouse-hamster, mouse-human, and even plant-animal
combinations. When the hybrid cells multiply in a culture dish, they tend to
lose some of their chromosomes. (In mouse-human cell hybrids, for example,
most of the human chromosomes are eventually lost; researchers are not yet
sure why this happens.) But some remain, and the genes they contain produce their characteristic proteins.

Some researchers have suggested using forms of cell fusion to introduce needed genes into cells of people with genetic diseases. But cell fusion also has a number of other valuable applications. Hybrid cells are helping researchers to map chromosomes—to determine which genes are found on which chromosomes. (For example, since the hybrids lose some chromosomes, if they no longer produce a particular protein, then the gene for that protein must have been on one of the lost chromosomes.) By studying hybrid cells, researchers can also find out more about how genes are turned on and off as an organism develops. In studies at the National Cancer Institute, for example, it was found that when mouse blood cells are fused with human blood cells, the hybrid produces both mouse and human hemoglobins. But if a mouse blood cell is fused with a human skin cell, it forms only mouse hemoglobin. Apparently the genes for human blood chemicals are turned off in skin cells and their hybrids.

Hybrid cells can also be used to produce valuable chemicals. The production of a type of fused cells called hybridomas is one of the most dynamic areas in the fast-growing biotechnology industry today.

THE
BIOTECHNOLOGY
INDUSTRY

Wall Street had never seen anything like it. On October 14, 1980, Genentech, a small company specializing in recombinant DNA research, offered a million shares for sale to the public. For weeks people had been begging their stockbrokers to get them some shares in the new offering. On the first day of trading, Genentech opened at $35 a share. In just twenty minutes of buying and selling on the stock exchange, the price of a share was up to $89! As the day went on, telephones in brokers' offices all over the country jangled constantly as people traded the new stock. The price fell a little. But at the end of the day Genentech was selling for $71.25 a share, and the founders of the company were instant millionaires.

One of the founders was Herbert Boyer, the University of California

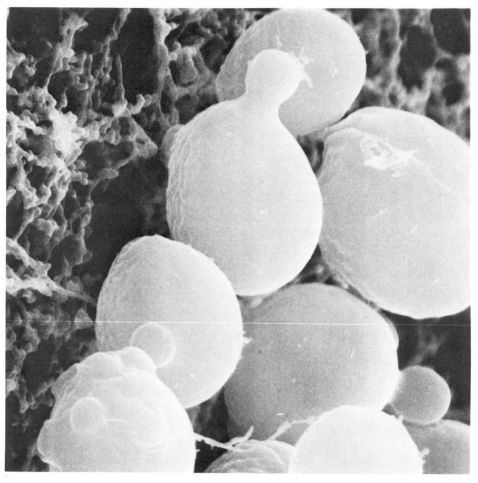

Yeast cells, magnified more than 14,000 times. The small buds on the cells will grow into full-size cells and separate from their parent cells.

biochemist who was one of the inventors of the recombinant DNA technique. (He owned 985,000 shares of the company.) Although Boyer was vice president of Genentech, he continued to spend most of his time at his university research laboratory. Except for buying an expensive car, the newly made millionaire did not make much change in his beer-and-blue-jeans life style.

Of course, his fortune was mostly on paper. And as months went by, investors lost their first-day frenzy. Genentech might be the darling of Wall Street, but the company had not yet produced a product to sell to the public. Meanwhile, other small genetic engineering firms went public. Large established companies in other fields (especially drug companies and chemical companies) started recombinant DNA divisions of their own.

Recombinant DNA companies and other firms specializing in the new biotechnologies have excited stockholders. But actually, biotechnology has been big business for a long time. Records of the ancient Babylonians show that they used yeast to make beer about 8000 years ago. About 2000 years later the Egyptians learned to use yeast to leaven bread. Ancient peoples also used microorganisms to make wine, yogurt, and cheese. These are all big industries today, but now microbes are also used to produce other valuable products, such as antibiotics and various chemicals. Yeasts and other microbes are even grown in huge vats to produce "single-cell protein," a nourishing addition to foods and animal feeds. In sewage-processing plants, microbes break down harmful chemicals in the wastes from homes and industry.

Recombinant DNA techniques can be used to improve the microbes used in various industrial processes—to make them turn out larger amounts and better products. They are also being used to produce many new substances with applications in medicine, industry, and agriculture.

Genentech's first recombinant DNA product was a hormone called somatostatin. This hormone works on cells to stop growth and helps to control the blood sugar level. It does not have any medical applications, although researchers think that chemically changed forms of somatostatin might be helpful in treating diabetes. But it was a fairly simple hormone to make, and it helped to prove that recombinant DNA techniques really worked.

In the past few years a number of other human hormones have been made by gene-splicing in bacteria and yeasts.

Human insulin produced by recombinant DNA is now being tested on diabetes patients. The tests will show whether the synthetic hormone is safe and effective. Researchers hope that it will be more effective than the cattle and pig insulins that diabetics use now. They think that the complications that cause some diabetics to go blind or to lose limbs because of poor cir-

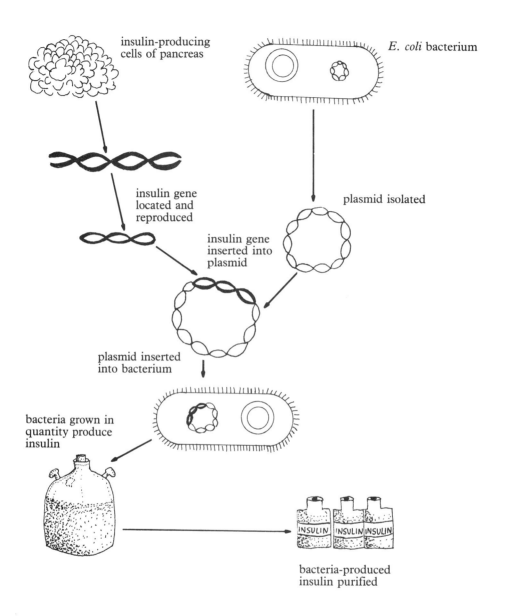

insulin-producing
cells of pancreas

E. coli bacterium

insulin gene
located and
reproduced

plasmid isolated

insulin gene
inserted into
plasmid

plasmid inserted
into bacterium

bacteria grown in
quantity produce
insulin

INSULIN INSULIN INSULIN

bacteria-produced
insulin purified

Recombinant DNA techniques are now being used to mass-produce
human insulin in bacteria.

culation may be due to allergic reactions to the animal insulins. If so, taking human insulin should prevent those harmful reactions.

Human growth hormone (HGH) is another recombinant DNA product now being tested. This hormone is produced by the pituitary gland. Some people do not have enough HGH, and so they do not grow as much as they should. As adults they are very short, perhaps even midgets. Doctors have been treating children with this problem with HGH taken from the pituitary glands of people who have died. (HGH from cattle and pigs does not work on humans.) The HGH from humans works, but it is extremely expensive. It takes 150 pituitary glands to get enough HGH to treat one child for a year! HGH produced with recombinant DNA will be much cheaper. When the new product is approved for sale, it will be possible to treat not only the most serious cases, but also children who would otherwise grow up to be just a little shorter than normal.

A number of companies are developing human interferon products. This substance is produced by cells that are attacked by a virus. Interferon helps to fight virus infections. Researchers have found that this substance also seems to help fight cancer. Clinical tests on cancer patients are under way now. Interferon is another substance that was very scarce and expensive before recombinant DNA methods of making it were developed. Within a few years it should be available for use not only against cancer, but also as a treatment for virus diseases like influenza, polio, and even the common cold.

Recombinant DNA techniques have been used to produce urokinase, a substance that dissolves blood clots. This could help people with heart disease. Recent studies have shown that if a clot-dissolving substance is injected into the blood vessels around the heart immediately after a heart attack, the clogged arteries can be opened up. If this is done promptly enough, the heart muscle recovers. The treatment may save many lives.

Recombinant DNA firms are working on a number of other substances with medical uses. Cetus Corporation, for example, is developing the production of Factor 8, a blood-clotting chemical. This substance can be used to treat people whose blood does not clot properly. Without it, they are in constant danger of bleeding to death from a minor cut or wound. Biotechnology companies are also working on producing endorphins, natural pain-

A research fermenter for the production of interferon.

killers produced in the brain that work even more effectively than drugs like morphine.

The new genetic research is also yielding improved vaccines for protection against diseases. The vaccines used now are usually prepared from killed viruses or from live viruses that have been changed slightly so that they build up the body's defenses but do not cause diseases. Unfortunately, people sometimes have bad reactions to these vaccines. People who are allergic

antibody-
producing cell

antibodies

antigens

A foreign substance (antigen) stimulates the body's antibody-producing cells to make large numbers of antibodies that attack the antigen. Some of the antibodies are kept as patterns, so that new antibody supplies can be produced quickly if the same antigen ever invades the body again.

to eggs, for example, may be made ill by flu vaccines that were grown in eggs. Or the vaccine viruses themselves may cause symptoms that are sometimes severe.

Vaccines work by stimulating the body to produce antibodies. These are proteins tailor-made to attack a particular invading virus. Once the body has been exposed to a virus, it keeps some of the antibodies "on file." Then if the person ever comes in contact with that virus again, a new supply of antibodies can be made quickly according to the pattern on file.

The antibodies are made to fit proteins or other substances on the outer coat of the virus. (These substances are called antigens.) Recombinant DNA researchers have isolated or made the genes for the coat proteins of various viruses and then produced the proteins. These antigens can be used as vaccines without the dangerous side effects of live viruses. They stimulate antibody production, but they cannot cause illness because they are only part of a virus, not a whole virus. Recombinant DNA vaccines have already been made to a form of foot-and-mouth disease, which is very dangerous to cattle;

The modern plant breeder can start with a mass of cells called a callus.
The cells grow on a culture dish and form leaves and roots. They are
transplanted first to flasks and then to pots with soil.

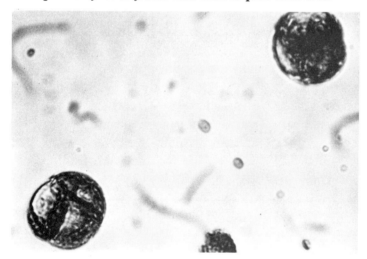

Cells from two different varieties of potato were stripped of their cell
walls and fused to form these hybrids. Researchers have grown whole
plants from single cells like these. They are studying the plants to see
which of the qualities of their parents they will show.

72

to hepatitis B, which can damage the human liver; and to a form of influenza.

An exciting prospect for the future is recombinant DNA production of enzymes that can be used to treat genetic diseases. More than two thousand different genetic diseases are known. Many of them involve the lack of a single key enzyme. Children born with phenylketonuria (PKU), for example, do not have the enzyme for breaking down the amino acid phenylalanine. This amino acid is found in many foods, and it is needed by the body to build proteins. But too much phenylalanine can poison the body and cause mental retardation. A simple test for PKU can be done shortly after birth; a piece of chemically treated paper placed in the baby's diaper reacts if there is too much phenylalanine in the baby's urine. Mental retardation can be prevented in PKU children by feeding them a special diet, containing practically no phenylalanine. But as they grow older, these children grow tired of their very limited diet. They long to be able to eat ice cream and hamburgers and pizza and all the other things that normal children eat. Some of them refuse to continue their special diet. Recombinant DNA might bring an answer to PKU and to other hereditary enzyme diseases by providing a source of the missing enzymes. Other techniques of genetic engineering may eventually give us an even better solution, when researchers learn how to introduce the genes for the missing enzymes into human cells and how to make them work in people's bodies.

The biotechnology companies are working on a number of products for agriculture. Animal growth hormones, now under development, can make cattle and other meat animals grow faster. Recombinant DNA techniques have produced a hormone called bovine prolactin, which increases cows' milk yield.

Plant breeders are looking forward to new, improved crop plants. Researchers have developed methods for introducing genes into plant cells. They are working out ways to give corn, wheat, and other cereals some genes from bean plants. These genes make the plants form nodules on their roots in which nitrogen-fixing bacteria can live. The bacteria get food materials from the plant, and they produce nitrogen compounds that the plant can use. With this natural "fertilizer," the plants can grow faster.

The new genetic techniques may also allow researchers to introduce

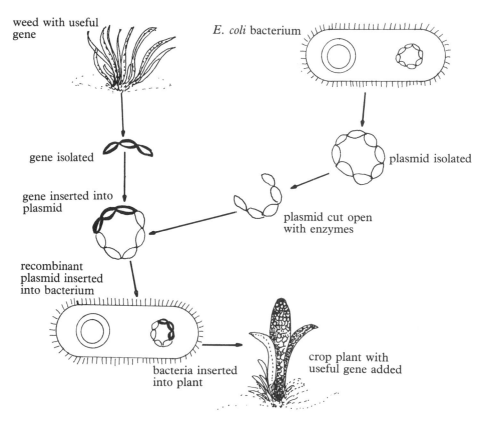

weed with useful gene

E. coli bacterium

gene isolated

plasmid isolated

gene inserted into plasmid

plasmid cut open with enzymes

recombinant plasmid inserted into bacterium

bacteria inserted into plant

crop plant with useful gene added

Recombinant DNA techniques may be used in plant genetic engineering. Bacteria can carry useful new genes into plants.

genes for resistance to diseases and to herbicides. These and other valuable characteristics would be introduced into plant *cells*, not whole plants. But researchers have already learned to grow whole plants from cells, and these plants would pass on their new genes to their offspring.

In addition to helping plant and animal growers, recombinant DNA research is improving microorganisms used in the food industry. By inserting specially selected genes, researchers are speeding up the process of breeding new and better strains of yeasts and other microbes for fermenting milk and grains and for making food substances such as sugars and amino acids. Some

74

of the essential amino acids are rather scarce in vegetable proteins. Microbe products can provide cheap additives to help make flour and other vegetable foods a better balanced diet.

In industry, microbes are already being used to produce valuable chemicals, from the ethylene glycol used in antifreeze to chemical intermediates for the manufacture of plastics. Recombinant DNA techniques are improving these microorganisms. They are also yielding new bacterial strains that can efficiently extract metals from low-grade ores and other bacteria that can help clean up pollution.

In June 1980 the U.S. Supreme Court made a landmark decision. The Court decided that unique new microorganisms could be patented. The decision was made in a case involving a bacterium specially bred by Ananda Chakrabarty, a researcher at General Electric Company. This "superbug" combined the characteristics of several other strains of bacteria that fed on petroleum chemicals. Chakrabarty's bacterium gobbled up more oil, faster, than the natural forms. Now Chakrabarty is working on breeding bacteria that can eat up hazardous chemicals in waste dumps. He has already produced one form that breaks down the herbicide 2,4,5-T, a major ingredient of Agent Orange, used in the Vietnam War to strip off the leaf cover in which enemies might hide. When this herbicide is applied to plants, it remains in the soil for a long time, and scientists fear that it might have harmful effects on humans. The new bacteria can clean up fields contaminated by 2,4,5-T.

The U.S. Navy is sponsoring recombinant DNA studies aimed at keeping the hulls of ships clean. Bacteria form a slime on the surfaces of ships in the water, and then barnacles, tube-worms, and other organisms attach themselves. These underwater hitchhikers create a drag that makes the ship burn more fuel. Shipowners must spend millions of dollars scraping and refinishing the hulls of their ships. Researchers plan to use recombinant DNA techniques to isolate the genes that make bacteria stick to surfaces in the water. Once they do that, they may be able to design substances to block those genes and stop the fouling process.

Imaginative projects like these are going on in laboratories all over the world. But recombinant DNA is only part of the story of the fast-growing biotechnology industry.

The technique of cell fusion provided the key to another active field of

General Electric Research & Development Center, Schenectady, NY

In a landmark decision, the U.S. Supreme Court allowed a patent to be issued for Ananda Chakrabarty's specially bred oil-eating microbes. The flasks contain petroleum and water. Microbes added to the flask on the right are beginning to digest the oil.

biotechnology. In 1975 British researcher Cesar Milstein and his colleagues were studying antibodies. These are the proteins that help to defend the body against invading microbes. Milstein's research team was making mice produce antibodies by injecting foreign substances into them. Antibody preparations from the blood of animals sensitized in this way are called antisera. They can be used to protect other animals (even humans) from the same disease. (Larger animals than mice are used to produce antisera such as tetanus toxoid for use on humans.)

Milstein wanted to study the antibody-producing cells from mouse spleens. But spleen cells do not usually grow well in cultures and soon die out. Cancer cells do grow well in cultures, though. So the British researchers got the idea of fusing spleen cells with a type of cancer cells called myeloma. Maybe the hybrid cells would produce antibodies like the spleen cells but grow as well as the myeloma cells. Sure enough, it worked. Milstein fused the two kinds of cells and got hardy, fast-growing hybrids that produced the same kinds of antibodies as the spleen cells. He named them "hybridomas."

The hybridomas turned out to be a mixture of cells producing various antibodies. Milstein isolated individual cells and cloned them, producing large numbers of offspring from each one. Each had the same heredity as its parent cell. When he tested the clones to see what kind of antibodies they formed, he found some that produced a single very pure antibody. These antibodies were named monoclonal antibodies.

Milstein sent samples of his hybridoma clones to many other researchers around the world. Soon research teams in many laboratories were working with hybridomas. They quickly realized that the very pure antibodies these cells produce could have many valuable practical uses.

For example, monoclonal antibodies can be used in tests to diagnose various diseases. Antibodies against a particular disease virus, for example, hepatitis virus, could be mixed with a sample of the patient's blood or tissues. If a reaction occurred, that would mean the patient had that disease. Medical laboratories have been running tests like that for a long time, but they used antisera produced by injecting the virus into an animal. These antisera are not pure. They are complicated mixtures that may contain only about 5% of the antibody wanted. So the results of the tests were sometimes not very

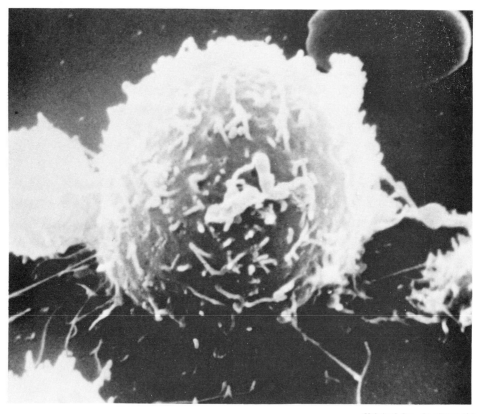

Hybritech Inc., San Diego, CA

Scanning electron micrograph of a cell fusion in progress. The small round cell is a spleen cell, and the large cell is a myeloma cell. The two cells are joining to form a hybridoma.

reliable. But with monoclonal antibodies, researchers know that the results will be exact, because these pure antibodies will react only with that particular virus. Monoclonal antibodies have already been produced against measles, influenza, rabies, and hepatitis viruses, the tetanus bacterium, and the microbe that causes malaria. An antibody test has been developed for heart attacks, using antibodies against a protein myosin, which is released into the blood from a damaged heart muscle.

One of the most exciting diagnostic uses of monoclonal antibodies is in tests for cancer. Researchers at Wistar Institute in Philadelphia and in vari-

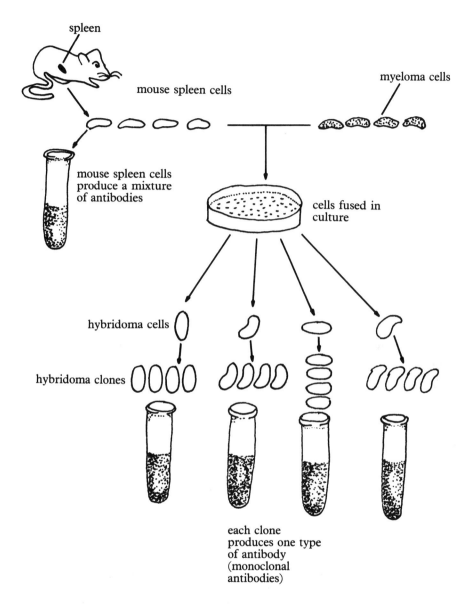

spleen

mouse spleen cells

myeloma cells

mouse spleen cells
produce a mixture
of antibodies

cells fused in
culture

hybridoma cells

hybridoma clones

each clone
produces one type
of antibody
(monoclonal
antibodies)

When antibody-producing spleen cells are fused with myeloma cancer cells, they form "hybidomas" that produce antibodies. Reproducing a single hybridoma cell yields very pure monoclonal antibodies.

ous other laboratories have found a number of specific antigens on the surface of cancer cells and produced monoclonal antibodies against them. Each kind of cancer has a number of different antigens, and different kinds of cancers have different sets of antigens. Some of the cancer antigens stay on the surface of the cancer cells, but others are shed into the blood. So two kinds of tests for cancer can be developed. Monoclonal antibodies against the antigens that cancer cells shed can be used for blood tests that show not only if a person has cancer, but what kind. Antibodies against the other kind of antigen can be injected into the body, "tagged" with a small amount of a radioactive substance. The antibodies are carried in the bloodstream until they reach the tumor, and then they latch onto the cancer cells. Then a radiation detector shows the doctor where in the body the cancer cells are. This kind of test can be very valuable, because cancer cells tend to break off from a tumor and travel through the body. Some of them settle down and multiply, forming new tumors. Monoclonal antibody tests can tell the doctor if this has happened.

Since monoclonal antibodies head straight for the cancer cells in the body, they may also provide a way to treat cancer. In fact, some researchers believe these antibodies may be the "magic bullets" against cancer that they have been searching for. By themselves, monoclonal antibodies can stop cancer cells from multiplying, but they do not usually kill them. But researchers are combining the antibodies with radioactive chemicals and poisons for more effective "bullets." The antibodies deliver the drugs that kill the cancer cells. Because the antibodies are so specific, these combinations kill *only* cancer cells and leave normal body cells alone. This is an important advantage. Many of the cancer treatments used now are so powerful that they kill normal cells along with the cancer cells.

The new antibodies can also help in organ transplants. The body's antibody system attacks not only invading microbes but also foreign tissues. If a heart or kidney is transplanted into the body, it may be rejected: antibodies attack the transplanted organ and kill its cells. Doctors use various drugs to knock out the antibody-forming system before a transplant operation. But then the patient has no defenses against infections. Now researchers are making highly specific monoclonal antibodies against the cells that stimulate production of the antibodies that cause rejection. If these cells are knocked out, the body will accept a transplanted organ but will still be able to fight disease germs and cancers.

Researchers are also finding monoclonal antibodies valuable tools. They can separate, purify, and fish out body chemicals that are present in very tiny amounts in complicated mixtures. This will be very helpful in recombinant DNA work. Monoclonal antibodies are already being used to isolate and purify interferon. The product is much purer and can be obtained in much larger amounts than was possible before.

Other kinds of cell cultures are being used in the new biotechnology industries. One of the monoclonal antibody companies, for example, is also growing human skin cells. These cells can then be "glued" together with collagen, the body's natural cement, to form sheets for skin grafts for burn victims.

Plant cell cultures are producing useful biochemicals. A drug company is manufacturing birth control pills from a hormone produced by the cells of yams. Drugs derived from plants, such as morphine and digitalis, could also be produced in plant cell cultures.

Researchers are working on variations of cloning to produce large numbers of identical offspring. These methods could have valuable applications in animal husbandry, producing "carbon copies" of prize cattle, egg-laying hens, and champion racehorses.

Other researchers have been raising mice with four or more parents. These "mosaic mice" are formed by mixing cells from very early embryos and then raising them in the womb of a foster mother. The mice inherit traits from all of their parents. Mixing cells from an embryo with pure white parents and cells from an embryo with pure black parents, for example, results in a mouse with patches of black-and-white fur. In one laboratory, researchers have even raised mice that have a tumor as one of their "parents." (Tumor cells were inserted into an embryo. Oddly enough, the mice are perfectly normal, and they are no more likely than usual to get cancer.) The various kinds of mixed-up mice are helping researchers to understand how traits are

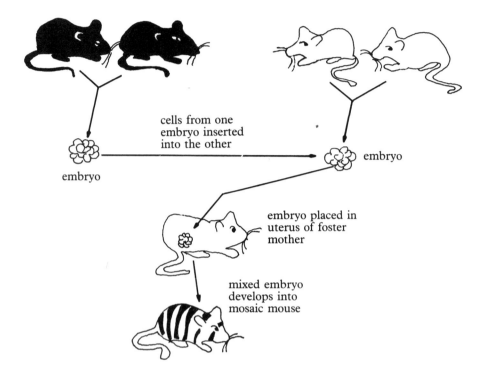

cells from one
embryo inserted
into the other

embryo

embryo

embryo placed in
uterus of foster
mother

mixed embryo
develops into
mosaic mouse

*When cells from one embryo are placed into another, and then the
hybrid embryo is grown in the uterus of a foster-mother mouse, the
"mosaic" mouse that results inherits characteristics from all four of the
original parents.*

inherited, but the techniques may also have applications some day in animal
breeding.

The advances of recent years in inventing new machines and instruments
for medicine and artificial body parts to replace natural organs damaged by
accident and disease have been spectacular. But many people believe that
the real biotechnology revolution going on today is on the microscopic level.
The new automated sequencers permit researchers to find out the structures
of proteins and nucleic acids. With gene machines and recombinant DNA
techniques they can make and study genes and proteins. The useful prod-

ucts that have already been made, like insulin and interferon, are only the beginning. Monoclonal antibodies and other results of cell fusion are also helping scientists to learn more about the body and promise to help doctors to heal their patients. Genetic engineers are learning to apply the discoveries of biotechnology to breed new and more useful microbes, plants, and animals. Some day we may even be able to use these tools and techniques to shape our own heredity. Some people find prospects like that frightening; others find them exciting. Either way, the biotechnology revolution is sure to change our future life.

FutureLife: Where Do We Go From Here?

Not many people can become famous just by being born. But that's what happened to a little English girl named Louise Brown. Her birth, in July of 1978, made headlines all over the world. She was the first "test-tube baby."

There aren't really any test tubes involved in the birth of a "test-tube baby," and Louise Brown spent nearly all of the time before her birth growing and developing inside her mother's body, just as other babies do. What made her so special is that the two tiny cells that formed her—a sperm cell from her father and an egg cell from her mother—did not join inside her mother's body. Instead, they were brought together by doctors in a laboratory culture dish, very much like the dishes recombinant DNA researchers use in their work.

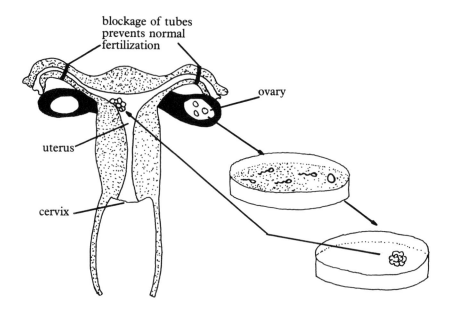

blockage of tubes
prevents normal
fertilization

ovary

uterus

cervix

Conception of a "test-tube baby." First the mother is given hormones that cause eggs in her ovary to ripen. Using a laparoscope, the surgeon removes eggs with a hollow needle. An egg is placed in a culture dish with blood serum and nutrients and fertilized with sperm. After a few days of development in the dish, the tiny embryo is placed in the mother's uterus, where it develops normally into a baby.

Drs. Patrick Steptoe and Robert Edwards are pioneers in the test-tube-baby field. Before their success with Louise Brown, they experimented for years, developing a number of special techniques. First the woman is given an injection of hormones, which make eggs in her ovaries ripen. Then the surgeon uses a thin instrument called a laparoscope to look at her ovary, find the ripened eggs, and gently draw them out through a hollow needle. The eggs are placed in a warmed culture dish with a special mix of blood serum and food materials, and sperm are added. After twelve hours or so, the eggs are examined under a microscope to see which ones have joined with sperm cells. These fertilized eggs are carefully transferred to another culture dish with a different food mixture. The eggs grow and divide for a few days. Meanwhile, the mother is receiving hormone injections—the same sort of hormones her own body would have been producing if she had just become pregnant in the normal way. When everything is ready, the doctors select one of the tiny embryos that seem to be developing normally and place it deep inside the mother's uterus. If all goes well, the embryo will soon burrow into the wall of the uterus and continue its development into a baby.

People all over the world were excited when the birth of the first test-tube baby was announced. But some asked: Why bother? With more than four billion people in the world and children crying with hunger, why do we need more ways to make babies? The answer is that although some babies are born to people who don't want them or can't support them, others who want very much to have a child are unable to do so. Louise Brown's parents had tried for nine years to have a child of their own. They had had no success, because the tubes inside Mrs. Brown's body which would normally carry ripe egg cells to her uterus each month were blocked. Many women suffer from this problem, and the test-tube-baby technique is a way to help them bear children of their own.

Since Louise Brown's birth, the British research team has helped a number of other parents to have test-tube babies, and medical teams in Australia and the United States are also working to help childless couples in this way. Their success raises some interesting questions for the future. For example, if a test-tube baby can be implanted in its own mother's womb, wouldn't it also be possible for a different woman to carry the child until birth, if its own mother is not able to do so? This has not yet been done with humans, but researchers often transfer eggs and embryos from one laboratory animal to another. (The offspring look like the parents who supplied the sperm and egg, not like their foster mother. The child of two white mice, for example, would have white fur even if it developed in the uterus of a black mouse.) Recently, zoo biologists have been using foster mothers in breeding rare wild animals threatened with extinction. At Utah State University, wild Sardinian sheep were born to a domestic sheep. At the Bronx Zoo in New York City, a dairy cow named Flossie gave birth to a baby gaur (a wild ox from India). In each case the embryos were taken out of the wild

At the Bronx Zoo, the cow Flossie is a surrogate mother who gave birth to a gaur calf and cares for it just as though it were really her own.

mother's uterus and transferred to the uterus of the domestic animal. After the babies were born, their foster mothers fed them and cared for them as though they were their own.

Probably human "surrogate mothers" of this kind are not too far in the future. But the idea of raising a test-tube baby in an artificial womb is a much longer-range prospect. It took years for Drs. Steptoe and Edwards to work out the right culture media for fertilizing eggs and growing them up to the stage of eight cells or so. At that stage an embryo is just a tiny mass of cells, too small to see properly without a microscope. As it grows and develops, an embryo becomes larger and more complex, and its needs multiply. It needs a continuing supply of food materials and a means of taking away its waste products. Inside the mother's uterus, a structure called the placenta provides for these needs through the lifeline of the umbilical cord. (Your "belly button" is the scar that remains to mark the spot where the stump of your umbilical cord shriveled and dropped off after birth.) The developing embryo gets both oxygen and food materials through the placenta from its mother's blood. Hormones and various chemicals from the foods its mother eats may also pass through the placenta. Researchers will need to find out more about how the natural placenta works before they can build an artificial womb.

A not quite so way-out possibility is the cloning of a baby from a single cell. Such a baby would have only one parent, instead of two. Genetically, the parent and child would be "identical twins," but they would not be the same age. In 1978 science writer David Rorvik published a book, *In His Image*, in which he claimed that a human child had already been cloned. He never revealed the name of the child or the father, and scientists generally agree that this book was science fiction, not science.

However, though the cloning of a human may not have happened yet, it *could* happen. Plant breeders frequently grow whole plants from just a twig, a bud, or even a single cell in a culture dish. Researchers have been cloning frogs for years. They take a cell from a tadpole—perhaps a skin cell or an intestinal cell—and using very delicate micro-tools, they insert that cell's nucleus into an egg cell whose own nucleus has been destroyed. The egg then grows into a frog that looks like the "parent" who supplied the nucleus, not like the frog who produced the egg.

In various ways, frogs' eggs are much easier to work with than those of humans and other mammals. For one thing, they are larger and sturdier; in addition, they normally develop outside the frog's body in the water of a pond or stream, so no special culture medium is needed to raise them. For awhile, some researchers doubted that cloning could ever be accomplished with mammals. But the difficulties have been overcome, and Swiss researchers recently reported the successful cloning of mice.

There is still a long way to go before a person can go to a doctor and say,

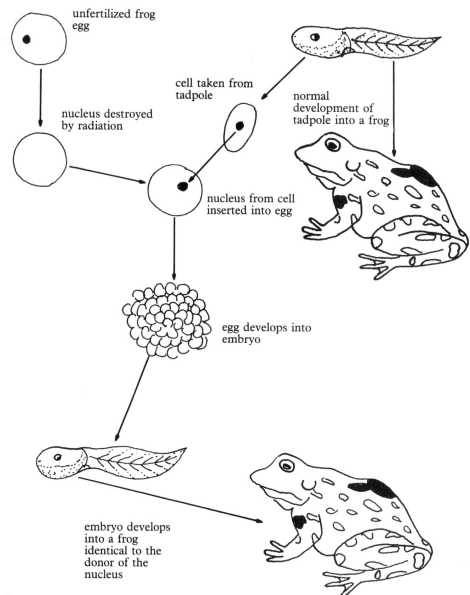

unfertilized frog
egg

nucleus destroyed
by radiation

cell taken from
tadpole

normal
development of
tadpole into a frog

nucleus from cell
inserted into egg

egg develops into
embryo

embryo develops
into a frog
identical to the
donor of the
nucleus

Frogs were used in the early cloning experiments. A nucleus taken from
a body cell of a tadpole was inserted into an egg. The frog that grew
from the egg was an "identical twin" of the tadpole from which the
nucleus was taken; it did not look like the parents of the egg.

"Clone me a child." For one thing, most of the successful cloning experiments with animals have used nuclei taken from cells of embryos. (A frog tadpole is an early stage of development, something like a human embryo or fetus.) Cells from adult animals seem to have lost the ability to spark the development of a new individual. Cloning attempts with cells from adults usually fail. And yet, if cloning is to have practical value, scientists will need to be able to use cells from adults. They might want to make duplicates of a champion racehorse, for example, or multiple copies of a prize milk cow or egg-laying hen. And some people will probably want to have children who are exact genetic duplicates of themselves.

Recombinant DNA researchers and molecular biologists may provide the knowledge needed to make cloning practical. They are mapping genes and learning how they work. They are learning about the "on-off switches" that turn genes on or stop them from working. Researchers believe that in an adult, all the genes for every stage of development are still there inside each cell, but most of them are turned off. If they can learn how to work the right on-off switches, they can turn some genes back on and make the cells younger.

This kind of knowledge will permit researchers to use adult cells for cloning offspring, but it may have many other applications as well. For example, a starfish that has lost an arm regrows a new one. If a flatworm is cut in half, the tail end grows a new head, and the head end grows a new tail. Many animals have these amazing natural powers of regeneration.

Right now humans can't grow a new leg or heart. But researchers in laboratories around the world are trying to solve the riddles of regeneration. They have made some progress. For example, researchers recently announced a promising new method to make a cut spinal cord regrow, using a hormone called TRH. This is an exciting advance. Once it was thought that if nerve cells were killed they could never be replaced. A person whose spinal cord was damaged would be doomed to a life of paralysis, unable to move legs or arms or even the whole body, depending on where the damage occurred. Other researchers are looking into the surprising possibility that a child whose fingertip has been cut off in an accident may regrow the missing fingertip, even without any treatment at all.

Today's regeneration experiments on humans are barely beginnings. It is a long way from fingertips and spinal cords to being able to grow back a whole hand or replace a faulty kidney with a new one grown from the patient's own cells. If we could do those things, it would be wonderful—far better than taking organs for transplants from other people or trying to build a replacement of metal and plastic. Genetic researchers think they have the best prospects of finding the answers to regeneration. If we could find the right genetic switches and learn how to work them, we could clone replacement organs from a person's own cells, or stimulate an amputee's stump to regrow a new hand.

Many researchers believe that aging is part of the genetic "program" of development: we grow old because the switches that work our genes are gradually turned off. If we could learn to turn the genes of growth and repair back on, we could keep people young indefinitely.

Biotechnology is also bringing us cures for diseases. Researchers are tailor-making drugs to work on a germ or a cancer cell and to leave the normal body cells alone. Genetic engineers are developing new, improved vaccines to protect us against diseases. Plant breeders are already breeding new strains of disease-resistant plants, and genetic engineers are developing new breeding techniques. These methods could be applied to animals, too, and even to humans. As genetic engineers work out ways to carry new genes into the body, they may be able to change the heredity not only of future generations, but also of people who are living now. With all these exciting new approaches, it is possible that we may eventually have cures for all diseases.

But if there are no more killer diseases and we develop ways to keep people from aging, what will people die from? Could we some day conquer death? This is an astonishing idea. People have been dying as long as there have been people on earth. Could it be possible that some day dying will no longer be necessary? Some scientists believe that this could happen. They think that aging and death are parts of the instructions in our genes, that we are "programmed" to die. They believe that when we learn to read all these instructions and to change them, we will be able to reprogram ourselves for an indefinite life.

Some researchers have speculated that future genetic engineers will be able to change our heredity in other important ways, too. Freedom from diseases and aging would be wonderful, but what other new characteristics might be useful for people to have? Increased intelligence, or perhaps more efficient ways of using the intelligence we already have, will probably be high on future genetic engineers' "shopping lists." Sports fans will probably be eager to develop increased strength and speed for future athletes. Perhaps we could also add some abilities that humans don't have now. How about gills like a fish, for example, so that we could live under water without diving equipment? Or perhaps skin with chlorophyll, so that we could make our own food as plants do and wouldn't need to eat? Space explorers might appreciate a built-in oxygen generator when they walk around on planets or moons that don't have the same kind of atmosphere as Earth's. Ideas like these may seem like way-out fantasies. And yet, the biotechnology revolution may make them possible some day.

The achievements of biotechnology are raising some startling possibilities for humanity's future and posing some intriguing—and sometimes troubling—questions.

New Questions
And
Hard Choices

In 1972 the United States Congress passed a bill to pay for kidney dialysis machines for all people who needed them. This semed like the right thing to do, because these machines can save the lives of people whose kidneys have failed. But nobody realized at the time how much the program would cost. Kidney dialysis is now costing American taxpayers a billion dollars a year, and the cost is going up. Some scientists have said that it would make more sense to spend more money on research to find the causes and cures of kidney diseases. Then people wouldn't need kidney machines anymore. But if we do that, it will take years to get results. What about people who need kidney dialysis *now*? Can we just let them die because it would cost too much to save them?

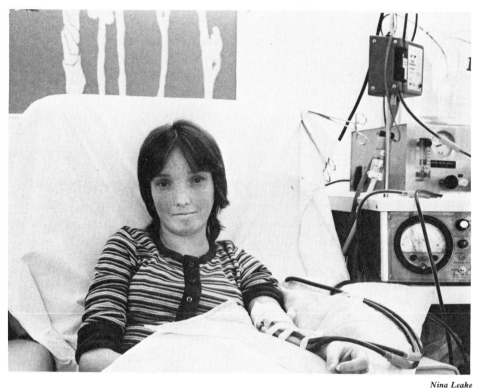

It would be tragic to deny kidney patients like this their life-saving dialysis. Yet can we afford the cost of a billion dollars each year?

What will happen when artificial hearts and artificial pancreases are ready to use in people? Would it be fair to save people with diseased kidneys and not have similar programs for people with heart disease or diabetes? Yet how can we possibly afford the costs of all these programs?

With new diagnostic tests, doctors can now tell before a baby is born whether it is suffering from various genetic diseases and birth defects. But we do not yet have cures for many of these conditions. Should such tests be conducted? Who should pay for them?

When recombinant DNA research was first starting, many people thought that it might be very dangerous. Some people wanted to stop the experiments completely. Are there some kinds of research that are too dangerous ever to be allowed? Or is the freedom of scientists to explore the unknown one of the basic freedoms of our society, which we should fight to defend? Who should decide whether the possible benefits are worth the possible risks? Scientists know the most about their work, so they would seem to be most qualified to make such decisions. But scientists are so interested and involved in their work that they might not be able to make fair decisions. The risks and benefits will affect the lives of average citizens, so they should have some say in what research is done. But can average citizens understand enough about technical matters to make wise decisions?

The tools and techniques used in medicine today, as well as the drugs that save people's lives, were developed through careful experiments on animals. Some people say that we do not have the right to use animals that way, for our own benefit. What about tests on humans? To test a new drug or technique, medical researchers must compare its effects with the old treatments to see if it is really better. But is it fair to give some sick people old treatments that are not very effective when the new ones might save their lives?

Research on brain waves and biofeedback has raised some disturbing questions. Scientists have found, for example, that there are centers for pleasure and pain in the brain. A rat with an electrode inserted into its brain pleasure center will continue to press a lever that stimulates the electrode until it drops from exhaustion. (In a similar experiment, a human volunteer continued to press a button to stimulate his pleasure center for more than an hour, until the researcher finally unplugged the electrode.) Could dictators someday use such techniques to control their subjects, by sending out radio waves to stimulate electrodes in their pain and pleasure centers? If the research could be misused in this way, should it be stopped now, for safety's sake? And yet, what of the benefits, such as brain pacemakers to permit people with damaged brains to live a more normal life?

What about cloning and genetic engineering? Might some future dictator use cloning to turn out armies of identical obedient soldiers by the thousands? Might a police state use genetic engineering to try to breed a nation of "ideal citizens"? Using genetic engineering to correct nature's mistakes—to prevent genetic diseases and birth defects—seems like a noble goal. Even this disturbs some people, though. (For one thing, more people with defective genes would survive and pass their defects on to their children. Our society would become very dependent on the use of genetic technology.) Yet some scientists would like to go further, to *improve* human beings. Who is to decide what traits are desirable? Who is to determine what direction the future evolution of the human race will take if we gain the means to remake ourselves?

The possibility of conquering diseases, aging, and perhaps even death raises new questions of its own. Even now when people are dying our world is facing problems of growing overpopulation and limited resources. Wouldn't these be even worse if people didn't die anymore? Yet if we have the means to save lives, wouldn't it be wrong not to use them? Perhaps if people knew they didn't have to die, they might care more about solving problems like pollution, hunger, and wars *now*, to make the world a better place to live in.

The biotechnology revolution is bringing us longer lives and better health. But it is also bringing some hard new problems for which there are no simple answers. Thoughtful scientists and other citizens are growing more and more concerned with bioethics: the study of how to decide what *should* be done in the troubling dilemmas of our modern world.

GLOSSARY

alloy: a combination of two or more substances (for example, two metals), which were melted together.

amino acid: the chemical building block of proteins. There are about twenty different kinds of amino acids in proteins.

antibiotic: a substance that kills microorganisms or stops them from growing. Many antibiotics, such as penicillin, are produced by molds or other microorganisms.

antibody: a complicated protein that is produced by the body's defense system to attack invading microbes or a foreign substance.

antigen: a foreign substance that stimulates the production of antibodies.

antiserum: a liquid fraction of the blood containing antibodies.

arthritis: a disease producing pain and stiffening of the joints.

bacterium: a type of microorganism. Some bacteria are rod-shaped, some round, some spiral. Some bacteria cause diseases; some (like *E. coli*) normally live harmlessly in the body; some live on decaying matter.

bionic: pertaining to the application of information about biological systems to the solution of engineering problems. One area of bionics applies the engineering solutions obtained to the development of artificial organs and body parts.

biotechnology: the application of biological information to engineering problems and the use of advanced instruments and devices to obtain information about biological systems or to treat medical problems.

cancer: a disease in which some body cells lose their normal controls and multiply wildly.

chromosomes: structures in the cell nucleus, each consisting of a long thread of DNA (carrying hereditary information), together with RNA and protein. Human cells have a set of forty-six chromosomes.

clone: a cell culture, organism, or group of organisms all derived from a single parent cell and sharing the same heredity.

cloning: reproducing new organisms from a single body cell, rather than through the combination of sex cells; also, the production of multiple copies of genes by recombinant DNA techniques.

99

computer: an electronic device that can perform calculations and can store, retrieve, and process information.

congenital: pertaining to a condition present at birth or one that arose during development before birth. (It may not be hereditary.)

culture: a mass of living material (such as cells, tissue) grown in a dish, flask, or test tube containing a nutrient medium (a fluid or solid with the food substances the cells need for growth).

cyclotron: a machine in which particles, such as protons or ions, are whirled around at high speeds and can be used to split atoms or produce other changes in matter.

cyst: a sac (usually filled with fluid) that forms abnormally in the body.

dialysis: a technique for removing some substances from a solution by passing the fluid over a membrane.

DNA: deoxyribonucleic acid; a substance found in the chromosomes, which carries hereditary information. Its building blocks are nucleotides.

electron: a tiny negatively charged particle that travels around the nucleus of an atom.

electronic: pertaining to devices utilizing the behavior and effects of electrons.

embryo: an early stage of development of an animal or plant. The human embryo stage covers about the first eight weeks after fertilization of the egg.

enzyme: a protein that acts as a catalyst in biological systems; that is, a substance that helps other chemicals to react. Enzymes are usually fantastically effective catalysts, and most are very specific (they work with just one or a very few chemicals).

epileptic seizure: convulsions (usually with a loss of consciousness) resulting from a disturbance of the electrical rhythms of the brain.

fetus: an unborn, developing animal. In humans, the fetal stage goes from about eight weeks after conception to birth.

fusion: a joining. Cell fusion is the joining of two cells to form a new single cell that can reproduce and pass on the combined heredity of its two cell ''parents.''

gene: the unit of heredity; a portion of DNA that directs the production of a specific protein or the expression of a particular trait.

genetic: hereditary, passed on from parents to offspring; pertaining to the genes.

heredity: the transmission of traits from parents to offspring.

hormones: chemical messengers produced in structures called endocrine glands and secreted into the blood; they control and coordinate body processes.

hybrid: an offspring of two animals or plants of different varieties or species; or a joining of two different substances (for example, DNA-RNA hybrids, or mouse-human hybrid cells).

hybridoma: a form of cell fusion in which a cell producing a particular biochemical is combined with a hardy, fast-growing cancer cell to form a culture that synthesizes large amounts of the biochemical.

infrared: heat radiation; a portion of the electromagnetic spectrum just beyond the red edge of the spectrum of visible light.

laser: a device that uses the vibrations of electrons to generate coherent electromagnetic radiation. A laser can produce a very narrow, intense beam of light that can heat, burn, or cut very precisely.

manic-depressive: a type of mental illness characterized by wild mood swings, from elation to depression.

microbe: a microorganism; a living organism too small to be seen without a microscope.

microsurgery: operating under a microscope or magnifier on very tiny structures, such as blood vessels, nerves, or individual cells.

molecule: the structural unit of a chemical compound; the smallest particle that still retains the properties of the compound.

monoclonal antibodies: very pure antibodies produced by a hybridoma.

myoelectric: pertaining to a device utilizing the electrical impulses of the body's own nerves and muscles.

Nobel Prizes: prizes awarded each year for important achievements in such fields as physiology and medicine, chemistry, and physics.

nucleotide: the building block of nucleic acids (DNA and RNA). There are four main types of nucleotides in a nucleic acid.

nucleus (plural: **nuclei**): a central structure in a cell, which contains the hereditary information and directs the cell's activities. Also, the central core of an atom, containing positively charged protons and electrically neutral neutrons.

plasmid: a small circular form of DNA found in some bacteria. It carries hereditary information outside the bacterial chromosome and can be transferred from one bacterium to another.

protein: a complex biochemical made of amino acids. Proteins are the main structural materials of cells (for example, hair is pure protein). Enzymes and some hormones are proteins.

radioactive: pertaining to unstable atoms that tend to break down, sending out particles and rays.

radioisotope: an unstable, radioactive form of a chemical element. (The same element may have some stable and some radioactive isotopes.)

radio waves: part of the electromagnetic spectrum between infrared radiation (heat) and audio frequencies (sound). Radio waves cannot be perceived by human senses, but they can be detected with instruments.

recombinant DNA: the transfer of genes from one species to another by cutting and splicing of DNA; gene splicing.

retina: the light-sensitive layer at the back of the eyeball.

reverse transcriptase: an enzyme that catalyzes the production of DNA according to an RNA pattern (the reverse of the usual process).

Rh reaction: an antigen-antibody reaction that may occur when a mother with Rh-negative blood gives birth to a baby whose blood is Rh-positive; the reaction may damage the baby's blood.

RNA: ribonucleic acid; a type of nucleic acid that is usually produced as a copy of a portion of the cell's DNA. Several forms of RNA work in protein synthesis.

schizophrenia: a form of mental illness characterized by loss of contact with the environment and disorders of feeling, thought, and actions.

stroke: the plugging or bursting of a blood vessel in the brain, resulting in damage to brain tissue; it may cause paralysis or death.

transcription: the production of an RNA copy of DNA.

translation: the production of protein according to the RNA blueprint.

tumor: an abnormal mass of tissue in the body. It may be cancerous.

vaccine: a preparation of live or killed microorganisms or an artificial antigen used to stimulate the production of antibodies that will provide protection against a disease.

virus: a very tiny microorganism consisting of a core of nucleic acid and an outer protein coat. A virus can reproduce only inside a cell host.

yeast: a kind of microorganism that produces fermentation. Yeasts are used in making bread and alcoholic beverages.

INDEX

Page numbers in italics indicate illustrations.